Antonius R B Ola

Activation of Silent Biosynthetic Pathways through Co-culture

Antonius R B Ola

Activation of Silent Biosynthetic Pathways through Co-culture

Südwestdeutscher Verlag für Hochschulschriften

Impressum / Imprint

Bibliografische Information der Deutschen Nationalbibliothek: Die Deutsche Nationalbibliothek verzeichnet diese Publikation in der Deutschen Nationalbibliografie; detaillierte bibliografische Daten sind im Internet über http://dnb.d-nb.de abrufbar.
Alle in diesem Buch genannten Marken und Produktnamen unterliegen warenzeichen-, marken- oder patentrechtlichem Schutz bzw. sind Warenzeichen oder eingetragene Warenzeichen der jeweiligen Inhaber. Die Wiedergabe von Marken, Produktnamen, Gebrauchsnamen, Handelsnamen, Warenbezeichnungen u.s.w. in diesem Werk berechtigt auch ohne besondere Kennzeichnung nicht zu der Annahme, dass solche Namen im Sinne der Warenzeichen- und Markenschutzgesetzgebung als frei zu betrachten wären und daher von jedermann benutzt werden dürften.

Bibliographic information published by the Deutsche Nationalbibliothek: The Deutsche Nationalbibliothek lists this publication in the Deutsche Nationalbibliografie; detailed bibliographic data are available in the Internet at http://dnb.d-nb.de.
Any brand names and product names mentioned in this book are subject to trademark, brand or patent protection and are trademarks or registered trademarks of their respective holders. The use of brand names, product names, common names, trade names, product descriptions etc. even without a particular marking in this works is in no way to be construed to mean that such names may be regarded as unrestricted in respect of trademark and brand protection legislation and could thus be used by anyone.

Coverbild / Cover image: www.ingimage.com

Verlag / Publisher:
Südwestdeutscher Verlag für Hochschulschriften
ist ein Imprint der / is a trademark of
OmniScriptum GmbH & Co. KG
Heinrich-Böcking-Str. 6-8, 66121 Saarbrücken, Deutschland / Germany
Email: info@svh-verlag.de

Herstellung: siehe letzte Seite /
Printed at: see last page
ISBN: 978-3-8381-3861-9

Zugl. / Approved by: Duesseldorf, Heinrich Heine University,Diss. 014

Copyright © 2014 OmniScriptum GmbH & Co. KG
Alle Rechte vorbehalten. / All rights reserved. Saarbrücken 2014

Aus dem Institut für Pharmazeutische Biologie und Biotechnologie
der Heinrich-Heine Universität Düsseldorf

Gedruckt mit der Genehmigung der
Mathematisch-Naturwissenschaftlichen Fakultät der
Heinrich-Heine-Universität Düsseldorf

Gedruckt mit der Unterstützung des
BMBF

Referent: Prof. Dr. Peter Proksch
Koreferent: Prof. Dr. Matthias U. Kassack

In memory for the coming of Alexander Pedro Ola

and Devoted to my beautiful angels

Maria Grace M. Aussiola and Maria Renny Praptiwi

CONTENTS

Abstract		5
Zusammenfassung		8
Chapter 1 General Introduction		11
1.1	An Historical Perspective of Natural Product	11
1.2	The Role of Natural Product from Plants in Drug Discovery	11
1.3	Endophytic Fungi	14
1.3.1	Endophytic Fungi as Producer of Important Drugs and Drugs Lead Structure Derived from Plants	15
1.3.1.1	Paclitaxel	15
1.3.1.2	Vincristine and Vinblastine	15
1.3.1.3	Camptothecin and its Analogues	15
1.3.1.4	Podophyllotoxin and its Analogues	15
1.4	The role of Fungal Natural Products in Drug Discovery	19
1.5	Strategies to Enhance the Chemical Diversity of Secondary Metabolites from the Endophytic Fungi	22
1.5.1	Molecular-Based Technique (Genetic Engineering)	22
1.5.2	Chromatin Remodelling	23
1.5.3	Simulation of Microbial Interactions through Coculture	23
1.6	Aims and Scope of the Study	25
Chapter 2 Inducing Secondary Metabolite Production through Coculture-An Ecological Perspective		26
Chapter 3 Absolute Configuration and Antibiotic Activity of Neosartorin from the Endophytic Fungus Aspergillus fumigatiaffinis		41
Chapter 4 Dihydroanthracenone Metabolites from the Endophytic Fungus *Diaporthe melonis* Isolated from *Anonna squamosa*		70
Chapter 5 Discussion		96
5.1 New Approaches for Activation of Silent Biosynthetic Pathways in Fungi		96
5.1.1 Induction of Silent Biosynthetic Pathways in the Endophytic Fungus *Fusarium tricinctum* through Coculture-An Ecological Perspective		96
5.1.2 Induction of Silent Biosynthetic Pathways in the Endophytic Fungus *Fusarium tricinctum* with Epigenetic Modifiers		98

5.2 Atropisomer Natural Products from the Endophytic Fungi *Aspergillus fumigatiaffinis* and *Diaporthe melonis*   100

5.2.1 The Influence of the Linkage and of the Substituents for the Biological Activity of Naturally Occurring Tetrahydroxanthone Atropisomers   100

5.2.2 The Influence of the Linkage and of the Substituents for the Biological Activity of Naturally Occurring Dihydroanthracenone Atropisomers   101

5.3 Biosynthetic Relationship of Polyketide Metabolites Isolated in This Study   102

List of Abbreviations   112

Reseach Contribution   114

Declaration of Academic Honesty/Erklärung   115

Acknowledgement   116

Abstract

Fungi are prolific and talented producers of structurally diverse bioactive metabolites and have yielded some of the most important products for the pharmaceutical industry such as penicillin. As almost all vascular plant species appear to be inhabited by endophytic fungi, the latter represent potential sources of novel natural products. Moreover, due to the world's urgent need for new antibiotics, chemotherapeutic agents and agrochemicals, current research activities focus on natural product drug discovery from endophytic fungi. However, it seems that many valuable bioactive compounds are overlooked and only a minority of bioactive metabolites is usually isolated when culturing the endophytic fungi under axenic conditions.

This dissertation is dealing with approaches to activate silent biosynthetic pathways and with the structure elucidation of fungal bioactive natural products focusing mainly on antibiotic and cytotoxic activities. As microbial interactions represent a significant factor for activation of silent biosynthetic pathways, an approach to maximize chemical diversity of endophytic fungi through coculture of fungi and bacteria was undertaken. In addition, inhibitors of histone deacetylases (HDAC) and DNA methyltransferases (DMATs) were also used to induce the secondary metabolite production of endophytic fungi. Moreover, investigation of bioactive metabolites from endophytic fungi under axenic culture conditions led to the isolation of the antibiotic compound neosartorin and of the new atropisomers diaporthemin A and B.

This dissertation consists of the following three parts that have already been published or submitted for publication:

Inducing secondary metabolites production by endophytic fungi through coculture with bacteria

Co-culturing of the fungal endophyte *Fusarium tricinctum* with the bacterium *Bacillus subtilis* 168 trpC2 on solid rice medium resulted in an up to seventy-eight fold increase of the accumulation of constitutively present secondary metabolites that included lateropyrone, cyclic depsipeptides of the enniatin type, and the lipopeptide fusaristatin A. In addition, four compounds including (–)-citreoisocoumarin as well as three new natural products were not present in discrete fungal and bacterial controls and only detected in the co-cultures. The new compounds were identified as macrocarpon C, 2-(carboxymethylamino) benzoic acid, and (−)-citreoisocoumarinol by analysis of the 1D, 2D NMR, and HRMS data. Enniatins B1 and A1, whose production was particularly enhanced, inhibited the growth of the co-cultivated *B. subtilis* strain with minimal inhibitory concentrations (MICs) of 16 and 8 µg/mL, respectively, and were also active against *Staphylococcus aureus, Streptococcus pneumoniae* and *Enterococcus faecalis* with MIC values in the range of 2-8 µg/mL. In addition, lateropyrone, which was constitutively present in *F. tricinctum* displayed good antibacterial activity against *B. subtilis, S. aureus, S. pneumoniae,* and *E. faecalis,* with MIC values ranging from 2 to 8 µg/mL. All antibacterially active compounds were equally effective against a multi-resistant clinical isolate of *S. aureus* and a susceptible reference strain of the same species.

Absolute configuration and antibiotic activity of neosartorin from the endophytic fungus Aspergillus fumigatiaffinis

Neosartorin was isolated from the endophytic fungus *Aspergillus fumigatiaffinis*. The absolute configuration of neosartorin, including both axial and central chirality elements, was established as (aR,5S,10R,5'S,6'S,10'R) for the first time on the basis of its electronic circular dichroism (ECD) spectra aided with TDDFT-ECD calculations. Neosartorin exhibited substantial antibacterial activity against a broad spectrum of Gram-positive bacterial species including staphylococci, streptococci, enterococci and *Bacillus subtilis* with minimal inhibitory concentrations in the range of 4 to 32 µg/mL. When the toxicity of neosartorin against eukaryotic cells was measured using a panel of different cancer cell lines such as HELA and BALB/3T3, the average IC_{50} values exceeded 32 µg/mL.

Abstract

Dihydroanthracenone metabolites from the endophytic fungus Diaporthe melonis isolated from Anonna squamosa

Chemical investigation of the endophytic fungus *Diaporthe melonis*, isolated from *Anonna squamosa*, yielded two new dihydroantharacenone atropodiastereomers, diaporthemins A and B, together with the known flavomannin-6,6'-di-*O*-methyl ether. The structures of the new compounds were established on the basis of extensive 1D and 2D NMR spectroscopy, as well as by high resolution mass spectrometry and by CD spectroscopy. The isolated compounds were tested for their antimicrobial activity against a multi-resistant clinical isolate of *Staphylococcus aureus* 25697, a susceptible reference strain of *S. aureus* ATCC 29213 and against *Streptococcus pneumoniae* ATCC 49619. Flavomannin-6,6'-di-*O*-methyl ether strongly inhibited *S. pneumonia* growth with a MIC value of 2 μg/mL and showed moderate activity against the *S. aureus* multi-resistant clinical isolate and susceptible reference strain (MIC 32 μg/mL), whereas diaporthemin A and B were not active against the tested strains.

Zusammenfassung

Pilze sind produktive und reichhaltige Quellen strukturell diverser, bioaktiver Metabolite und haben einige der wichtigsten Produkte der pharmazeutischen Industrie hervorgebracht, wie z.B. das Penicillin. Da in fast allen vaskulären Pflanzen endophytische Pilze leben, stellen diese eine potentielle Quelle neuer Naturstoffe dar. Zusätzlich steigt das Interesse an der Entdeckung neuer Naturprodukte aus endophytischen Pilzen aufgrund steigender Resistenzen weltweit und dem daraus resultierenden Bedarf an neuen Antibiotika, Chemotherapeutika und agrochemischen Mitteln. Allerdings scheint es so zu sein, dass viele bioaktive Sekundärstoffe übersehen und nur ein Bruchteil isoliert werden, wenn der endophytische Pilz axenisch kultiviert wird.

Diese Arbeit beschäftigt sich daher unter anderem mit Versuchen, die stillen Biosynthesewege zu aktivieren sowie mit der Strukturaufklärung von bioaktiven Sekundärstoffen aus Pilzen mit Fokus auf antibiotisch und zytotoxisch wirksamen Verbindungen. Da mikrobielle Interaktionen wichtig für die Aktivierung stiller Biosynthesewege sind, wurde der Versuch unternommen, durch die Co-Kultivierung von Pilzen und Bakterien die chemische Diversität von endophytischen Pilzen zu maximieren. Zusätzlich wurden epigenetische Modulatoren in Form von Inhibitoren der Histondeacetylasen (HDAC) und DNS-Methyltransferasen (DMATs) eingesetzt, um die Sekundärstoffproduktion endophytischer Pilze zu induzieren. Letztlich führte die Untersuchung von Monokulturen endophytischer Pilze zur Isolierung des antibiotisch wirksamen Neosartorins und der neuen Atropisomere Diaporthemin A und B.

Diese Arbeit beinhaltet folgende drei Abschnitte, welche bereits veröffentlicht oder zur Publikation eingereicht wurden:

Induktion der Produktion von Sekundärmetaboliten in endophytischen Pilzen durch Co-Kultivierung mit Bakterien

Co-Kultivierung des Pilz-Endophyten *Fusarium tricinctum* mit dem Bakterium *Bacillus subtilis* 168 trpC2 auf festem Reismedium führte zu einer bis zu 78-fachen Zunahme der Akkumulation an konstitutiv vorhandenen Sekundärmetaboliten wie beispielsweise Lateropyron, zyklischen Depsipeptiden des Enniatin-Typs und des Lipopeptids Fusaristatin A.

Weiterhin waren vier Verbindungen, darunter (–)-Citreoisocumarin sowie drei neue Naturstoffe lediglich in der Co-Kultivierung detektierbar, während sie in den separaten Pilz- und Bakterienkontrollen nicht auffindbar waren. Die neuen Verbindungen wurden mittels 1- und 2D-NMR-Spektroskopie und hochauflösender Massenspektrometrie (HRMS) als Macrocarpon C, 2-Carboxymethylaminobenzoesäure und (–)-Citreoisocoumarinol identifiziert. Enniatin B1 und A1, deren Produktion besonders gesteigert wurde, inhibitierten das Wachstum des co-kultivierten *B. subtilis* Stammes mit einer minimalen Hemmkonzentration (MHK) von 16 und 8 µg/mL und waren weiterhin aktiv gegen *Staphylococcus aureus*, *Streptococcus pneumoniae* und *Entherococcus faecalis* mit MHK-Werten im Bereich von 2-8 µg/mL. Außerdem zeigte Lateropyron, welches konstitutiv in *F. tricinctum* zu finden war, gute antibakterielle Eigenschaften gegen *B. subtilis*, *S. aureus*, *S. pneumoniae* und *E. faecalis* mit MHK-Werten zwischen 2-8 µg/mL. Alle antibakteriell wirksamen Verbindungen waren gleichwertig effektiv gegen multi-resistente klinische Isolate von *S. aureus* und einen empfindlichen Referenzstamm der gleichen Art.

Absolute Konfiguration und antibiotische Aktivität von Neosartorin aus dem endophytischen Pilz Aspergillus fumigatiaffinis

Neosartorin wurde aus dem endophytischen Pilz *Aspergillus fumigatiaffinis* isoliert. Die absolute Konfiguration von Neosartorin einschließlich axialer und zentral chiraler Elemente wurde mittels elektronischem Circulardichroismus (ECD) und unterstützenden TDDFT-ECD-Kalkulationen erstmalig als (aR,5S,10R,5'S,6'S,10'R) ermittelt. Neosartorin zeigte beachtliche antibakterielle Eigenschaften gegen ein breites Spektrum an grampositiven Bakterienspezies wie Staphylokokken, Streptokokken, Enterokokken und *Bacillus subtilis* mit MHKs im Bereich von 4-32 µg/mL. Bei Toxizitätsmessungen von Neosartorin gegen eukaryotische Zellen wie beispielsweise HELA und BALB/3T3 überstiegen die Durchschnittswerte 32 µg/mL.

Zusammenfassung

Dihydroanthracenon-Metaboliten aus dem endophytischen Pilz Diaporthe melonis isoliert aus Anonna squamosa

Eine chemische Untersuchung des endophytischen Pilzes *Diaporthe melonis*, isoliert aus *Anonna squamosa*, führten zu den zwei neuen Dihydroanthracenon-Atropdiastereomeren Diaporthemin A und B zusammen mit dem bekannten Flavomannin-6,6'-di-*O*-methylether. Die Strukturen der neuen Verbindungen wurden anhand weitreichender 1- und 2D-NMR-Spektroskopie, hochauflösender Massenspektrometrie und CD-Spektroskopie etabliert. Die isolierten Verbindungen wurden auf ihre antimikrobielle Aktivität gegen multi-resistente klinische Isolate von *Staphylococcus aureus* 25697, einem empfindlichen Referenzstamm *S. aureus* ATCC 29213 und gegen *Streptococcus pneumoniae* ATCC 49619 getestet. Flavomannin-6,6'-di-*O*-methylether hemmte das Wachstum von *S. pneumonia* mit einer MHK von 2 µg/mL sehr stark und zeigte moderate Aktivität gegen das mult-resistente klinische Isolat und den empfindlichen Stamm von *S. aureus* (MHK 32 µg/mL), während Diaporthemin A und B gegen die getesteten Stämme inaktiv waren.

Chapter 1

General Introduction

1.1 An Historical Perspective of Natural Product

Nature has been a source of medicinal products for millennia with many useful drugs developed especially from plant sources. The earliest records of natural products date from around 2600 BC, documenting the uses of approximately 1000 plant-derived substances in Mesopotamia. These include oils from *Cupressus sempervirens* (cypress), *Cedrus* species (cedar) and *Commiphora* species (myrrh) which are still used today to treat coughs, colds and imflammation (Borchardt, 2002). The best known Egyptian medicine records, the "Ebers Papyrus" (dating from 1500 BC) documented over 700 plant-based drugs (Cragg and Newman, 2013). The Chinese Materia Medica has been extensively documented over the centuries, with the first record dating from about 1100 B.C. (Wu Shi Er Bing Fang, containing 52 prescriptions), followed by works such as the Shennong Herbal (~100 B.C.; 365 drugs), and the Tang Herbal (659 A.D.; 850 drugs) (Cragg and Newman, 2005 and 2013). Likewise, documentation of the Indian Ayurvedic system was dated from about 1000 B.C. (Susruta and Charaka), and this system formed the basis for the primary text of Tibetan Medicine, Gyu-zhi (Four Tantras) translated from Sanskrit during the eighth century A.D. (Borchardt, 2003).

In the ancient Western world, the Greeks contributed substantially to the rational development of the use of herbal drugs. The Greek physician, Dioscorides, (100 A.D.), recorded the collection, storage and the use of medicinal herbs, while the Greek philosopher and natural scientist, Theophrastus (~300 B.C.) dealt with medicinal herbs (Dias *et al.*, 2012).

1.2 The Role of Natural Product from Plants in Drug Discovery

Medicinal application of natural products can be traced back several millennia in human history. As mentioned above, plants have formed the basis for traditional medicine systems of several indigenous therapeutic systems such as Traditional Chinese medicine (TCM) and Ayurveda. In traditional medicine, natural products, mainly botanical drugs, were and are used on the basis of empirical experiences rather than pharmacological knowledge. Their ethno pharmacological properties have been used as a primary source of medicines for early drug discovery. A survey from World Health Organization (WHO) indicated that approximately 65% of the population of the world predominantly relied on plant-derived traditional medicines for their primary health care and 80% of 122 plant-derived pure compounds as drugs were used for the same or related ethno medical purposes and were derived from only 94 plant species (Fabricant et al., 2001).

The knowledge associated with traditional medicine (complementary or alternative herbal products) has promoted further investigations of medicinal plants as potential medicines and has led to the isolation of many natural products that have become well known pharmaceuticals. Probably, the best example of ethno medicine's role in guiding drug discovery and development is that of the antimalarial drugs, particularly quinine and artemisinin (Klayman, 1985 and Wongsrichanala et al., 2002). The antimalarial drug quinine was first isolated from the bark of *Cinchona* species (e.g., *C. offincinalis*) that had long been used by indigenous groups in Amazon region for the treatment of fever. Quinine is the basis for the synthesis of the commonly used antimalarial drugs chloroquine and mefloquine (Cragg and Newman, 2013). As the resistance to both these drugs in many tropical regions were observed (Mita et al., 2009), a search for a new drug against malaria led to the isolation of artemisinin from *Artemisia annua* (Quinhaosu) using data from ancient texts in Traditional Chinese Medicine (Wongsrichanala et al., 2002). The plant was originally used in Traditional Chinese Medicine as a remedy for chills and fever. Artemisinin has also been the basis for the synthesis of several analogues such as arteether, OZ277 and the dimeric analogue for improving its activity and utility (Vennerstrom et al., 2004). Artheether was introduced in 2000 and was proved as antimalarial drugs as well (Dias et al., 2012).

Quinine   Artemisinin   Chloroquine   Mefloquine

Besides Malaria, plants have a long standing story of use in the treatment of cancer. Some of the best known plant-derived anticancer drugs in clinical use are vinca alkaloids, vinblastine and vincristine isolated from the Madagascar periwinkle tree *Catharanthus roseus* (Johnson *et al.*, 1963; Cragg and Newmann, 2013). Vinflunine (Javlor®) is a novel fluorinated vinca alkaloid approved as antitumor agent for the treatment of bladder cancer in 2010 (Newman and Cragg, 2012). The other examples of original-plant drugs for the treatment of cancer are two clinically active agents, etoposide and teniposide, which are semisynthetic derivatives of the natural product epipodophyllotoxin (Mishra and Tiwari, 2011). Epipodophyllotoxin is an isomer of podophyllotoxin, which was isolated as the active antitumor agent from the roots of various species of the genus *Podophyllum* (Liu and Jiao, 2006). These plants have long been used by early American and Asian cultures for the treatment of skin cancers and warts (Liu *et al.*, 2007).

The most widely used breast cancer drug, paclitaxel (Taxol®) was isolated from the bark of *Taxus brevifolia* (Pacific Yew) (Wani *et al.*, 1971). The discovery in 1980 of its mode of action through promotion of the assembly of tubulin into microtubules (Schiff and Horwitz, 1980) was a key milestone in the lengthy development process as approved clinical drugs for the treatment against ovarian cancer in 1992 and against breast cancer in 1994. Since then it has become a blockbuster drug with annual sales over $1 billion and with the current annual demand in the range of 100-200 kg per annum (i.e., 50.000 treatments/year) (Dias *et al.*, 2012). Paclitaxel (Taxol®) was present in limited quantities from natural sources and was able to be produced semisynthetically (Nicolaou *et al.*, 1994). The presence of baccatin III in higher quantities from the needles of *T. brevifolia* allowed the efficient transformation into paclitaxel (Danishefsky *et al.*, 1996). Currently, paclitaxel is produced through plant cell fermentation technology using a specific Taxus cell line propagated in aqueous medium with the endophytic fungus *Penicillium raistrickii* (Bristol-Myers Squibb, 2004). Paclitaxel then became the most exciting plant-derived anticancer drug. The latest anticancer agent approved in USA was the third taxane cabazitaxel (Jevtana®) in 2010 for the treatment of hormone-refractory prostate cancer (Mishra and Tiwari, 2011; Newman and Cragg, 2012).

Vinblastine   Vincristine   Vinflunine

Baccatin III   Paclitaxel

1.3 Endophytic Fungi

The word endophyte means "in the plant" (endon Gr. = within, phyton = plant). Endophytes can be defined as fungi that reside in internal tissues of living plants without causing any immediate overt negative effects (Bacon and White, 2000), but may turn to pathogenic during host senescence (Rodrigguez and Redman, 2008). Endophytic fungi are a polyphyletic group of primarily ascomycetous fungi, whereas basidiomycetes, deuteromycetes and oomycetes are rarely found (Saikonnen et al., 1998 and Arnold et al., 2007). Any organ of the host can be colonized. Endophytic fungi can be transmitted from one generation to the next through the tissue of host seed, but the majority of endophytes are horizontally transmitted to their host plants through airborne spores (Hartley and Gange, 2009). Endophytic fungi are thought to interact mutualistically with their host plants mainly by increasing host resistance to herbivores and have been termed "acquired plant defenses" (Faeth and Fagan, 2002 and Soliman et al., 2013). They also play important roles in ecosystem processes such as decomposition and nutrient cycling, and have beneficial symbiotic relationships with roots of many plants (Sun *et al.*, 2011).

Among 300 thousand higher plant species, each individual plant is host to one or even several hundreds of strains of endophytes (Strobel and Daisy, 2003). As almost all vascular plant species appear to be inhabited by endophytic bacteria or fungi, these represent important components of microbial diversity. These wide ranges of microbial diversity are potential sources of novel natural products for exploitation in medicine, agriculture and industry as microbial diversity imply chemical diversity due to the constant chemical innovation that exists in ecosystems. Moreover, due to the world's urgent need for new antibiotics, chemotherapeutic agents and agrochemicals to cope with the growing medicinal and environmental problems facing mankind, growing interest is taken into the research on the chemistry of endophytic fungi.

1.3.1 Endophytic Fungi as Producer of Important Drugs and Drugs Lead Structure Derived from Plants

During the long period of co-evolution, a friendly relationship was gradually set up between endophytic fungi and their host plant and this is believed to shape natural product patterns of endophytic fungi. Several studies have reported the biosynthesis of important plant secondary metabolites by endophytic fungi residing in the plants that were originally known as the source of such metabolites.

1.3.1.1 Paclitaxel

The discovery of the paclitaxel producing endophytic fungus *Taxomyces andreanae* isolated from the original source of this important anti-cancer drug *Taxus brevifolia* has evoked the interest in endophytes as potential source for therapeutic agents. Paclitaxel has activity against a broad band of tumor types including prostate, ovarian, breast, lung, head and neck cancers as well as advanced forms of Kaposi's sarcoma (Aly *et al.*, 2013). However, paclitaxel (Taxol®) is present in limited quantities from natural sources as 1 kg of paclitaxel requires 10,000 kg of *Taxus* bark (Strobel *et al.*, 1996). The scarcity of paclitaxel and negative ecological impact encouraged scientist to develop alternative source and approach for producing this valuable bioactive compound. Up to now, at least 20 genera of endophytic fungi (*Alternaria, Aspergillus, Botryodiplodia, Botrytis, Cladosporium, Ectostroma,*

Fusarium, Metarhizium, Monochaetia, Pestalotia, Pestalotiopsis, Phyllosticta, Pithomyces, Taxonomyces and *Tubercularia*) have been reported to produce paclitaxel and its analogues such as baccatin III (Zhao *et al.,* 2011). Among the genera, the fungal strain *Metarhizium anisopliae* isolated from *Taxus chinensis* produced the highest amount of paclitaxel around 846.1 µg/L (Liu et al., 2009) followed by *Cladosporium cladosporioides* (isolated from *Taxus media*) around 800 µg/L (Zhang *et al.,* 2009).

The hosts of paclitaxel-producing fungi are mainly *Taxus* species such as *T. baccata, T. cuspidate, T. media* and *T. yunanensis*. Of special interest are reports on fungi residing in plants other than *Taxus* species such as *Cardiospermum halicacabum* (Spindaceae), *Citrus medica* (Rutaceae), *Cupressus* sp. (Cupressaceae), *Gingkgo biloba* (Ginkgoaceae), *Hibiscus rosa-sinensis* (Malavaceae), *Podocarpus* sp. (Podocarpaceae), *Taxodium distichum* (Taxodiaceae), *Terminalia arjuna* (Combretaceae), *Torreya grandifolia* (Taxaceae) and *Wollemia nobilis* (Araucairaceae) (Zhao *et al.,* 2011). The fungal strain *Phyllossticta citricarpa* from *Citrus medica* (Kumaran *et al.,* 2008) and *Phyllossticta dioscoreae* from *Hibiscus rosa-sinensis* (Kumaran *et al.,* 2009) produced 265 and 298 µg/L paclitaxel, respectively. However, paclitaxel yields from fungal isolates are still too low for sustained industrial productions.

1.3.1.2 Vinblastine and Vincristine

Vinblastine and vincristine are also two well-known anticancer agents clinically used for certain types of cancer such as leukemia, lymphoma, breast and lung cancers. Vinblastine was first reported in 1998 from the endophytic fungus *Alternaria* sp isolated from *Catharanthus roseus* (Guo *et al.,* 1998). Later, an endophytic fungus *Fusarium oxysporum* (Zhang *et al.,* 2000) and an identified fungus from from *C. roseus* (Apocynaceae) (Yang *et al.,* 2004) were also reported to produce vincristine.

1.3.1.3 Camptothecin and its Analogues

The naturally occurring topoisomerase I inhibitor camptothecin, a pentacyclic alkaloid is the parent compound of the semi-synthetic derivatives irinotecan and topotecan that are clinically used against ovarian, small lung and refractory ovarian cancers (Srivastava *et al.,* 2005). It was first isolated from the wood of *Camptotheca acuminata* (Nysaceae) in 1966 (Wall *et al.,*

1966). Later, camptothecin was also found systematically in unrelated plant families such as Icanacea (*Nothapodytes nimmoniana*, *Pyrenacantha kalineana*, *Merrilliodendron megacrapum*, *Apodytes dimidiate*), Apocynaceae (*Ervatamia heyneana*), Rubiaceae (*Ophiorrhiza pumila*, *O. mungos*), and Gelsemiaceae (*Mostuea brunonis*) (Ramesta et al., 2008; Wink, 2008; Shweta et al., 2010). The wide distribution of this alkaloid in various plant families was hypothesized originally to originate from the endophytic fungi via infection of the respective plants or gene transfer (Wink, 2008). Camptothecin was first reported from an endophytic fungus *Entrophospora infrequens* isolated from the plant *Nothapodytes foetida* which has been known also to produce camptothecin (Puri et al., 2005). From the seeds of the same plant *Nothapodytes foetida*, an endophytic fungus *Neuropsora* sp was reported also to be able to produce camptothecin (Rehmen et al., 2008). More recently, camptothecin and two of its anolugues 9-methoxy camptothecin and 10-hydroxy camptothecin with improved aqueous solubility but similar potency to camptothecin were discovered from the endophytic fungus *Fusarium solani* obtained from *Camptotheca acuminata* (Kusari et al., 2009). From this plant, Min and Wang had found an unidentified fungal strain XK001 that could produce 10-hydroxycamptothecin with a yield of 677 µg/L. In 2010, two endophytic *Fusarium solani* strains MTCC9667 and MTCC9668 from *Apodytes dimidiate* (*Icacinaceae*) were reported to produce camptothecin and two of its analogues 9-methoxy camptothecin and 10-hydroxy camptothecin with yields in the range of 8.2-53.6 µg/100 g of dry cell mass (Shweta et al., 2010).

Camptothecin $R_1=R_2=H$
9-methoxycamptothecin $R_1=OMe, R_2=H$
10-hydroxycamptohecin $R_1=R_2=OH$

Topotecan

Irinotecan

1.3.1.4 Podophyllotoxin and its Analogues

Podophyllotoxin is the precursor to clinically used anti-cancer drugs such as etoposide and teniposide which act as topoisomerase inhibitors. Podophyllotoxin exerts its antineoplastic action by preventing the polymerization and assembly of tubulin into the mitotic-spindle microtubules, thus arresting the cell cycle at mitosis (Guerram *et al.*, 2012). The semi synthetic derivatives, however, show a different mechanism of action as potent inhibitors of topoisomerase II (Wigley, 1995). The difference in the mode of action was related to the presence of the bulky glucoside moiety which is absent in podophyllotoxin (Wigley, 1995). It mainly occurs in plants from the genera *Diphylleia*, *Dysosma*, *Juniperus* (also called *Sabina*) and *Podophyllum* (also called *Sinopodophyllum*) (Zhao *et al.*, 2011). At present, the major supply of podophyllotoxin is from natural *Podophyllum* species (Zhao *et al.*, 2011). Due to over-exploitation, *Podophyllum* plants have been declared as endangered species. Therefore, an alternative resources and strategies for efficiently producing this valuable bioactive compound should be developed.

Several endophytic fungi have been reported to accumulate podophyllotoxin. *Phialocephala fortiniii* and *Trametes hirsuta* isolated from *Sinopodophyllum peltatum* and *S. hexandrum*, respectively, had been reported to produce podophyllotoxin with a yield ranging from 0.5-189 μg/L (Eyberger *et al.*, 2006 and Puri *et al.*, 2006). Other podophyllotoxin producing endophytic fungi include *Alternaria* sp from *Sinopodophyllum hexandrum* (Cao *et al.*, 2007) and *Fusarium oxysporum* from *Sabina recurva* (Kour *et al.*, 2008). A recent study reported the accumulation of podophyllotoxin by the endophytic *Fusarium solani* from the root of *S. hexandrum* with a yield of 29.0 μg/g dried weight (Nadeem *et al.*, 2012). Moreover, deoxypodophylltoxin as an anticancer pro-drug was found in the endophytic *Aspergillus fumigatus* isolated from *Juniperus communis*. Deoxypodophylltoxin is not only a possible precursor of podophyllotoxin, but it shows also broad therapeutic efficacy against a variety of malignancies (Kusari *et al.*, 2009).

Podophyllotoxin R=OH
Deoxypodophyllotoxin R=H

Etoposide

Teniposide

1.4 The Role of Fungal Natural Products in Drug Discovery

Fungi are a prolific source of structurally diverse bioactive metabolites and have yielded some of the most important products of the pharmaceutical industry. Undoubtedly, one of the most famous natural product discoveries from microorganism is penicillin G isolated from the fungus *Penicillium notatum* almost 84 years ago (1929) by Alexander Fleming. The first publication of the broad therapeutic use of this agent in the 1940s had led to a new era in medicine "the Golden Age of Antibiotic" (Fleming, 1929; Abraham *et al.*, 1941; Flemming, 1980). Since then, there was a worldwide endeavor to discover new antibiotics and bioactive natural products from microorganisms and many pharmaceutical companies were motivated to start sampling and screening large collections of fungal strains, especially for antibiotics. In the following years, some further antimicrobial agent isolated from fungi and included griseofulvin (Grove *et al.*,1952) and cephalosporin C (Newman and Abraham, 1955).[61] Penicillin together with cephalosporin are β-lactam antibiotics which became the world's blockbuster drugs with total sales of about 15 billion dollars worldwide in 2002 and represented around 65% of the world antibiotic market (Elander, 2003). However, the rising emergence of bacterial resistance to the β-lactam antibiotics was well documented after their clinical use for more than 60 years against the bacterial infections (Gould *et al.*, 2012 and Stefani *et al.*, 2012)

In terms of antimycotic agents, griseofulvin (Fulvicin®, Grifulvin V®, Grisovin FP®, Gristatin®) was one of the first antifungal natural products isolated from the fungus *Penicillium griseofulvum* (Grove *et al.*, 1952) and is commonly used to treat fungal infections of the skin, hair, and nails. The fungistatic drug acts by binding to tubulin, thus interfering with microtubule function and inhibiting mitosis (Huber and Gottlieb, 1968; Richardson and Warnock, 2003). Griseofulvin strongly inhibits mitosis in fungal cells but inhibition of mammalian cells is only weak. Griseofulvin inhibits cell-cycle progression at the prometaphase/anaphase of mitosis in human cells by suppressing spindle microtubule dynamics, the same mode of action as observed for other antimitotic drugs, such as vinca alkaloids and taxanes (Ho *et al.*, 2001; Panda *et al.*, 2005). However, griseofulvin is relatively safe at the clinical doses used because it mainly accumulates in the keratin layers of the skin, where it exerts its action by inhibiting fungal mitosis at concentrations which are significantly lower than those required to inhibit mitosis in human cells (Panda *et al.*, 2005).

Recently, echinocandin B and pneumocandin B, isolated from *Aspergillus rugulovalvus* and *Glarea lozoyensis*, respectively, were the lead compounds and templates for the semisynthetic antifungal drugs caspofungin (Cancidas®) and anidulafungin (Eraxis®) (Debono et al., 1995; Mishra and Tiwari, 2011; Newman and Cragg, 2012). Echinocandin that inhibits the synthesis of glucan in cell wall is known to exert limited toxicity in humans due to lack of biological target 1,3 –ß-glucan synthesis and thus, is significant against invasive infections by Candida species. More recently, another metabolite from the fungus *Aspergillus fumigatus*, fumagillin (McCowen et al., 1951) (Flisint®, Sanofi-Aventis) was approved as antiparasitic agent (Mishra and Tiwari, 2011). In September 2005, France approved fumagillin against intestinal microsporidiosis, a disease caused by the spore forming unicellular parasite *Enterocytozoon bieneusi*, and causing chronic diarrhea in immune compromised patients (Lanternier et al., 2009).

Another group of fungal derived drugs are the antilipidemic statin compounds. These include mevastatin and lovastatin (Mevacor®), from *Penicillium citrinum* and *Aspergillus terreus*, respectively, (Endo et al., 1976 and Bukland et al., 1989) or synthetic analogue compounds such as the major selling synthetic statins (lipitor®, crestor® and livalo®) (Aly et al., 2011).[73] Statins are the most potent cholesterol-lowering agents available. As high blood cholesterol levels contribute to the incidence of coronary heart disease, statins are of potential value in treating high-risk coronary patients (Lewington et al., 2007). Two lipid-regulating drugs of this class, atorvastatin (lipitor®) and simvastatin (Zocor®), feature prominently in the top ten drugs by cost reflecting the widespread implementation of clinical guidelines and recommendations relating to coronary heart disease. Statins lower cholesterol by reversible competitive inhibition of the rate-limiting enzyme HMG-CoA reductase in the mevalonate pathway of cholesterol biosynthesis, thus reducing total and low-density lipoprotein cholesterol levels (Alberts, 1988 and 1990; Chao et al., 1991).

In the area of immunopharmacology, the discovery of cyclosporine A (also known as ciclosporin or cylosporin) from the fungus *Tolypocladium inflatum* in 1971 was an important step as the substance prevents rejection after organ or tissue transplantations (Survase et al., 2011; Bushley et al., 2012). Improvements in the field of organ transplantations and treatment of autoimmune diseases are still in progress with the discovery that also known substances such as the fungal metabolite mycophenolic acid (Myfortic®) possess immunosuppressive activities (Elbarbry and Shoker, 2007). Cyclosporine exhibits, in addition to its potent immunosuppressant activity pronounced antiviral activity. Therefore, it furthermore served as a model for the design of substances like Debio-025, a potential antiviral drug that has successfully passed clinical trials (Coelmont *et al.*, 2009; Mishra and Tiwari, 2011).

Besides in the area of medicine, the roles of fungal natural products in drug discovery are also important in agriculture area for plant protection as demonstrated by the discovery of the strobilurins. The first naturally occurring strobilurins, strobilurin A and B, were isolated from *Strobilurus tenacellus*. Strobilurins showed no antibacterial activities but inhibited the growth of a variety of phytopathogenic fungi in very low concentrations of 10^{-7}–10^{-8} M (Anke, 1995) and thus became the lead compounds for various synthetic fungicidals such as azoxystrobin from Zeneca (sold as Amistar® for cereals, Quadris® for grape vines, and Heritage® for turf), kresoxim-methyl and pyraclostrobin from BASF, trifloxystrobin by Novartis (Flint®), picoxystrobin from Syngenta and etominostrobin from Shionogi (Aly *et al.*, 2011). The demand for new highly effective agricultural agents to control farm pests and pathogens is enormous, and partly arises from the removal of synthetic compounds from the market due to their toxicity towards the environment.

Penicillin G

Cephalosporin

Griseofulvin

Mevastatin R1=H, R2=Me, R3=H
Lovastatin R1=Me, R2=Me, R3=H
Simvastatin R1=Me, R2=R3=Me

1.5 Strategies to Enhance the Chemical Diversity of Secondary Metabolites from the Endophytic Fungi

1.5.1 Molecular-Based Technique (Genetic Engineering)

One of the successful strategies to induce silent biosynthetic pathways is based on molecular biology techniques such as the generation of gene "knock outs", promoter exchange, overexpression of transcription factors or other pleiotropic regulators (Brakhage and Schroeckh, 2011). Until now, the preferable method of activation (silent gene) clusters is overexpression of a pathway-specific transcription factor as this method allows increased expression of a whole cluster at the same time. This method was first reported from overexpression of the *Aspergillus nidulans* transcription factor gene *apdR* (part of the silent *apd* gene cluster containing a central hybrid PKS–NRPS gene) which led to the production and identification of aspyridones that had never been isolated before from *A. nidulans* (Bergmann et al., 2007).

Another method that has also been successfully applied is the exchange of promoters for biosynthesis genes or transcription factor genes with inducible promoters such as the alcohol dehydrogenase A gene promoter (P*alcA*). For example, the exchange of promoter *acvA* from *A. nidulans* by homologous recombination with the alcohol dehydrogenase 1 promoter *PalcA* led to an increase penicillin production up to 30 fold (Kennedy and Turner, 1996). A similar experiment has been described for the brevianamide F synthetase in *Aspergillus fumigatus* which led to production of compounds that had not been observed before for this strain (Maiya et al., 2006).

Manipulation of a global regulator gene by overexpression or deletion has been also used to induce secondary metabolite production and accumulation in fungi. A prominent example was the overexpression of *laeA*, which led to increased production of various secondary metabolites in several fungi, such as penicillin in *A. nidulans* (Bok and Keller, 2004) and *P. chrysogenum* (Kosalkova et al., 2009) and aflatoxin in *A. flavus* (Kale et al., 2009). A further example is the deletion of the single *A. nidulans* sumoylation gene, *sumO*, which led to reduced amounts of austinol and dehydroaustinol but increased amounts of asperthecin (Szewczyk et al., 2008) and the deletion of an *N*-acetyltransferase gene in *A. nidulans* which led to the formation of pheofungins, which represented novel metabolites related to pheomelanins in red hair (Scherlach et al., 2011).

1.5.2 Chromatin Remodelling

Chromatin is the complex of DNA and the associated histone proteins. Histone proteins are little spheres that DNA is wrapped around. If the way how DNA is wrapped around the histones changes, gene expression can change as well. Because histone modification has an important role in the activation of several gene clusters, it is possible to activate silent gene clusters by treating fungi with the addition of chromatin-modulators such as inhibitors of histone acetyltransferases (HAT), histone deacetylases (HDACs) or DNA methyltransferases (DMATs). This technique does not require strain-dependent genetic manipulation and can thus be applied to any fungal strain.

The use of small molecule histone deacetylase (HDAC) or DNA methyltransferase (DMATs) inhibitors to perturb the fungal secondary biosynthetic machinery has been well documented. Commercially available compounds such as 5-azacytidine (DMAT) and suberoylanilide hydroxamic acid (SAHA-HDAC inhibitor) have been used in several laboratories to activate silent biosynthetic pathways. For example, culturing the fungus *Diatrype* sp in the presence of the DMAT inhibitor 5-azacytidine was able to trigger a significant change in the fungus metabolic profile, resulting in the *de novo* production of two polyketides lunalides A and B (Wiliams *et al.*, 2008). Moreover, treatment of the fungus *Cladosporium cladosporioides* with 5-azacytidine stimulated the accumulation of several oxylipins. In contrast, addition of suberoylanilide hydroxamic acid (SAHA) to the culture of *C. cladosporioides* induced the production of two new perylenequinones, cladochromes F and G (Williams *et al.*, 2008). Another example that demonstrates the potential of this strategy was the isolation of nygerone A from *A. niger* when cultured with SAHA (Henrikson *et al.*, 2009). More recently, addition of the histone deacetylase inhibitor and the DNA methyltransferase inhibitor to the culture medium of an entomopathogenic fungus, *Isaria tenuipes*, greatly enhanced the accumulation of secondary metabolites of the fungus and led to the isolation of the new compound tenuipyrone (Asai *et al.*, 2013).

1.5.3 Simulation of Microbial Interactions through Coculture

Bacteria and fungi co-inhabit a wide variety of habitats such as soil, water or the living tissues of higher plants where they may be present as endophytes (Strobel *et al.*, 2004; Aly *et al.*, 2011). One important interaction of fungi and bacteria includes competition for limited nutrients, which is known as a major ecological factor that triggers natural product biosynthesis and accumulation in prokaryotes and eukaryotes alike (Brakhage and Schroeckh, 2011; Scherlach and Hertweck, 2009). Co-culturing of different microbes rather than maintaining axenic cultures as is usually practiced in microbiology forces direct interactions that may enhance the accumulation of constitutively present natural products (Oh *et al.*, 2007; Schroeckh *et al.*, 2009; Nützman *et al.*, 2011) or may trigger the expression of silent biosynthetic pathways yielding new compounds (Cueto *et al.*, 2001; Oh *et al.*, 2005; Zuck *et al.*, 2011). Thus, interspecies crosstalk leading to chemical diversity provides a conceptual framework for the discovery of novel fungal compounds and the elucidation of their role as info chemicals. Therefore, a physiological condition that is very likely to activate silent gene clusters is the co-cultivation (co-culture) of microorganisms which interact or communicate with each other.

Several co-culture studies have been reported in the past which resulted either in an enhanced production of known metabolites, or in the discovery of new molecules. Co-cultivation of two marine microorganisms, the fungus *Emericella* sp. and the actinomycete *Salinispora arenicola* enhanced the expression of the emericellamide biosynthesis gene cluster 100-fold (Oh *et al.*, 2007). The production of new cytotoxic diterpenes, libertellenones A-D, was induced when the marine-derived fungus *Libertella* sp. was co-cultivated with the marine bacterium CNJ-328 (Oh *et al.*, 2005). Another example was the production of a new antibiotic, pestalone, during a co-culture of the marine fungus, *Pestalotia* sp., with an unknown α-proteobacterium, CNJ-328. (Cueto *et al.*, 2001).

Recent findings showed that intimate physical interaction during co-culture of the fungus *Aspergillus nidulans* and the bacterium *Streptomyces rapamycinicus* triggered the expression of gene clusters for orselinic acid and lecanoric acid that are kept silent under standard laboratory conditions (Schroeckh *et al.*, 2009). The same bacterium *S. rapamycinicus* was currently also reported to induce a previously silent polyketide synthase pathway in the important human pathogenic fungus *Aspergillus fumigatus*, and led to the discovery of a previously unreported prenylated polyketide (König *et al.*, 2013). It was found that the bacterium *S. rapamycinicus* specifically alters gene expression of *A. nidulans* and *A. fumigatus* by inducing a histone modification (Nützmann *et al.*, 2011 and König *et al.*, 2013).

1.6 Aims and Scope of the Study

Fungi have provided modern medicine with drugs and drug lead structures to combat major global diseases. The main aim of this study was to develop new methods for activation silent biosynthetic pathways in endophytic fungi based on an ecological perspective (publication 1) together with the isolation and structure elucidation of biologically active secondary metabolites (publications 2, 3 and 4).

The metabolites isolated from the single culture (publications 2, 3 and 4) and co-culture (publication 1) were then investigated for their pharmaceutical potential mainly focusing on antibiotic and cytotoxic activity.

In order to isolate the secondary metabolites, the endophytic fungi were grown on solid rice media for 3-4 weeks (publication 2, 3 and 4). In co-culture, the endophytic fungus was added to the rice solid media containing bacteria that had been incubated for 4 and 6 days at 37^0 C (publication 1). After reaching the stationary phase of growth, the cultures were harvested and subsequently extracted with organic solvents followed by fractionation and purification using various chromatographic techniques. Fractions and pure compounds were analyzed by HPLC-DAD and LC-MS for their purity, UV spectra, molecular weights and fragmentation patterns. The structures of the metabolites were elucidated using state of the art one and two dimensional NMR techniques.

Chapter 2 shows the main significant result of this study dealing with approaches to activate silent biosynthetic pathways based on an ecological perspective through co-culture of the fungus *Fusarium tricinctum* with the bacterium *Bacillus subtilis*. In this study, we found an up to seventy-eight fold increase of the accumulation of constitutively present secondary metabolites and also the production of four metabolites including the induced known (–) citreoisocoumarin as well as three new natural products. It is also interesting to note that enniatins B1 and A1, whose production was particularly enhanced, inhibited the growth of the co-cultivated *B. subtilis* strain with minimal inhibitory concentrations (MICs) of 16 and 8 µg/mL, respectively.

In chapter 3, we report for the first time the absolute configuration of neosartorin together with its antibacterial activity against multi resistant clinical isolates. Chapter 4 deals with the structure elucidation of two new atropisomeric diaporthemin A and B from the fungus *Diaporthe melonis*.

Chapter 2

Inducing Secondary Metabolite Production through Coculture-An Ecological Perspective

Published in "Journal of Natural Products"

Impact Factor: 3.285,

The overall contribution to the paper: 80% of the first author. The first author involved to all laboratory works as well as the manuscript preparation.

Coculture

pubs.acs.org/jnp

Inducing Secondary Metabolite Production by the Endophytic Fungus *Fusarium tricinctum* through Coculture with *Bacillus subtilis*

Antonius R. B. Ola,[†,‡] Dhana Thomy,[†] Daowan Lai,[*,†] Heike Brötz-Oesterhelt,[†] and Peter Proksch[*,†]

[†]Institut für Pharmazeutische Biologie und Biotechnologie, Heinrich-Heine-Universität Düsseldorf, Universitätsstrasse 1, Geb. 26.23, 40225 Düsseldorf, Germany

[‡]Department of Chemistry, Faculty of Science and Engineering, Nusa Cendana University, Jalan Adisucipto Penfui, 85001 Kupang, Indonesia

Supporting Information

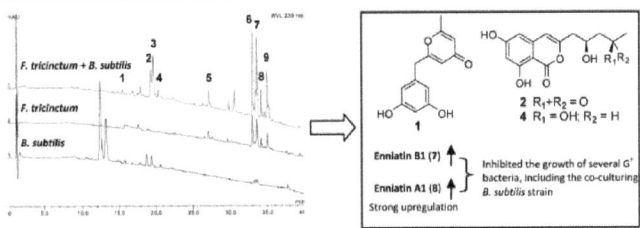

ABSTRACT: Cocuituring the fungal endophyte *Fusarium tricinctum* with the bacterium *Bacillus subtilis* 168 trpC2 on solid rice medium resulted in an up to 78-fold increase in the accumulation in constitutively present secondary metabolites that included lateropyrone (5), cyclic depsipeptides of the enniatin type (6−8), and the lipopeptide fusaristatin A (9). In addition, four compounds (1−4) including (−)-citreoisocoumarin (2) as well as three new natural products (1, 3, and 4) were not present in discrete fungal and bacterial controls and only detected in the cocultures. The new compounds were identified as macrocarpon C (1), 2-(carboxymethylamino)benzoic acid (3), and (−)-citreoisocoumarinol (4) by analysis of the 1D and 2D NMR and HRMS data. Enniatins B1 (7) and A1 (8), whose production was particularly enhanced, inhibited the growth of the cocultivated *B. subtilis* strain with minimal inhibitory concentrations (MICs) of 16 and 8 μg/mL, respectively, and were also active against *Staphylococcus aureus*, *Streptococcus pneumoniae*, and *Enterococcus faecalis* with MIC values in the range 2−8 μg/mL. In addition, lateropyrone (5), which was constitutively present in *F. tricinctum*, displayed good antibacterial activity against *B. subtilis*, *S. aureus*, *S. pneumoniae*, and *E. faecalis*, with MIC values ranging from 2 to 8 μg/mL. All active compounds were equally effective against a multiresistant clinical isolate of *S. aureus* and a susceptible reference strain of the same species.

Bacteria and fungi co-inhabit a wide variety of habitats such as soil, water, or the living tissues of higher plants, where they may be present as endophytes.[1,2] One important interaction of fungi and bacteria includes competition for limited nutrients, which is known as a major ecological factor that triggers natural product biosynthesis and accumulation in prokaryotes and eukaryotes alike.[3,4] Coculturing of different microbes rather than maintaining axenic cultures as is usually practiced in microbiology forces direct interactions that may enhance the accumulation of constitutively present natural products[5−7] or may trigger the expression of silent biosynthetic pathways yielding new compounds.[8,9] Recent findings show that intimate physical interaction during coculture of the fungus *Aspergillus nidulans* and the bacterium *Streptomyces rapamycinicus* triggered the expression of gene clusters for orselinic acid and lecanoric acid that are kept silent under standard laboratory conditions.[6,7] Therefore, an induction of silent biosynthetic pathways by coculturing of two or more different microbial strains would seem to greatly expand the possibility to discover new bioactive molecules.

Several coculture studies have been reported in the past that resulted either in an enhanced production of known metabolites or in the discovery of new molecules. Cocultivation of two marine microorganisms, the fungus *Emericella* sp. and the actinomycete *Salinispora arenicola*, enhanced the expression of the emericellamide biosynthesis gene cluster 100-fold.[5] The production of new cytotoxic diterpenes, libertellenones A−D, was induced when the marine-derived fungus *Libertella* sp. was cocultivated with the marine bacterium CNJ-328.[8] Another example was the production of a new antibiotic, pestalone, from a coculture of the marine fungus *Pestalotia* sp. with an unknown α-proteobacterium, CNJ-328.[9]

Received: July 19, 2013
Published: October 31, 2013

© 2013 American Chemical Society and American Society of Pharmacognosy

Coculture

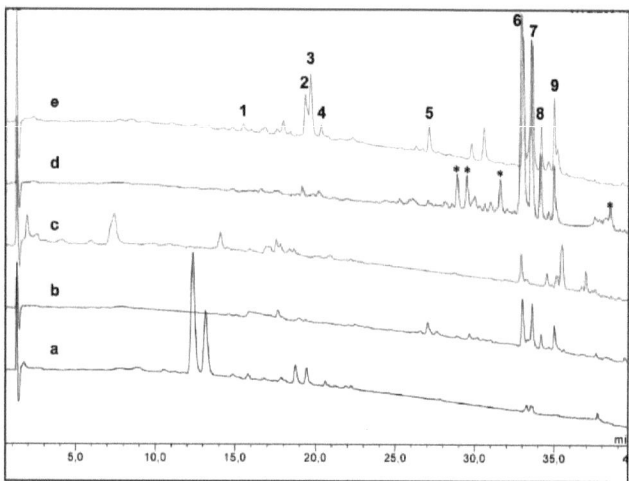

Figure 1. HPLC chromatograms of the EtOAc extracts from coculture experiments detected at UV 235 nm: (a) control of *B. subtilis*, (b) control of *F. tricinctum*, (c) control of *S. lividans*, (d) coculture of *F. tricinctum* with *S. lividans* (*unidentified metabolites), (e) coculture of *F. tricinctum* with *B. subtilis*.

In this paper, we report the influence of *Bacillus subtilis* 168 trpC2 on the secondary metabolism of *Fusarium tricinctum* during cocultivation. We found an up to 78-fold increase in the accumulation of constitutively present fungal products such as lateropyrone (5), cyclic depsipeptides of the enniatin type (6−8), and the lipopeptide fusaristatin A (9) in the presence of *B. subtilis* compared to axenic cultures of *F. tricinctum*. In addition, four metabolites (1−4) including three new natural products, 1, 3, and 4, were identified only in cocultures (Figure 1). Interestingly, when *F. tricinctum* was cocultured with a second bacterium (*Streptomyces lividans*), compounds 1−4 were not detected, whereas the accumulation of other, yet unidentified compounds was induced that were lacking in cocultures of *F. tricinctum* and *B. subtilis*. This finding suggests that the response of *F. tricinctum* toward different bacteria may be a specific rather than a general reaction toward prokaryotes.

■ RESULTS AND DISCUSSION

The strain of *F. tricinctum* used in this study had been isolated previously as an endophytic fungus from the plant *Aristolochia paucinervis*.[10] The bacterial species, *B. subtilis*, that was used for cocultivation with *F. tricinctum* is known from different habitats including soil but was also isolated from higher plants.[11−13] Thus, *F. tricinctum* and *B. subtilis* can both be considered as endophytes and hence may even be suspected to co-occur in plants even though no direct evidence for this hypothesis has been brought forward so far. *F. tricinctum* is known to produce bioactive cyclic peptides of the enniatin type, which are among the characteristic constituents of this species.[10,14] When *F. tricinctum* was cultured axenically on solid rice medium, average yields of the leading enniatin derivatives per culture flask were 8.78 mg for enniatin B (6), 2.44 mg for enniatin B1 (7), and 0.28 mg for enniatin A1 (8) (Table 1). Coculturing of *F.*

Table 1. Yield of Induced Metabolites per Flask during Coculture of *F. tricinctum* and *B. subtilis* ($n = 8$) vs Axenic Controls of *F. tricinctum* ($n = 8$)

compound	control (mg)	coculture (mg)	increase (fold)
1	n.d.[a]	0.71 ± 0.19	
2	n.d.	3.86 ± 1.52	
3	n.d.	15.99 ± 1.83	
4	n.d.	0.36 ± 0.07	
5	3.40 ± 0.94	4.85 ± 2.32	1.4
6	8.78 ± 1.54	79.58 ± 21.85	9.0
7	2.44 ± 1.32	88.54 ± 24.42	36.3
8	0.28 ± 0.15	21.85 ± 8.35	78.0
9	10.05 ± 1.24	79.40 ± 28.54	7.9

[a] n.d.: not detected.

tricinctum and B. subtilis resulted in a strong enhancement of enniatin accumulation. Average production of enniatins per flask in cocultures reached 79.58 mg for enniatin B (6), 88.54 mg for enniatin B1 (7), and 21.85 mg for enniatin A1 (8), which accounted for an up to 78-fold increase compared to controls (Table 1). A similar trend was observed with regard to the lipopeptide fusaristatin A (9),[15] which increased from 10.05 mg in axenic F. tricinctum to 79.40 mg per flask in cocultures. For lateropyrone (5),[16,17] however, which is another typical constituent of F. tricinctum, no clear induction was detected (Table 1), indicating that the effects of cocultivation are not uniform for all fungal compounds.

In addition to the strong increase of constitutively present compounds (with the notable exception of lateropyrone), four further compounds (1–4) that were only detected in cocultures of F. tricinctum and B. subtilis were isolated during this study. These compounds included three new natural products (1, 3, and 4).

Compound 1 was isolated as a colorless powder, whose molecular formula was determined as $C_{13}H_{12}O_4$ by HRESIMS (m/z 233.0874 [M + H]⁺), requiring 8 degrees of unsaturation. The ¹H NMR spectrum showed the presence of three aromatic protons at δ_H 6.30 (2H, d, J = 2.1 Hz, H-9/H-13) and 6.28 (1H, t, J = 2.1 Hz, H-11), two olefinic protons at δ_H 5.99 (1H, d, H-5) and 5.98 (1H, m, H-3), one methylene group at δ_H 3.69 (2H, s, H$_2$-7), and one methyl group at δ_H 2.22 (3H, d, J = 0.5 Hz, H$_3$-14). The ¹³C NMR spectrum displayed 13 carbon signals (Table 2) including seven protonated carbon signals at

Table 2. ¹H and ¹³C NMR Data and HMBC Correlations of 1 (acetone-d_6)

position	δ_C, type	δ_H mult. (J in Hz)	HMBC correlations (H→C)
2	166.5, C		
3	114.3, CH	5.98 m	C-2, C-14, C-5, C-4w[a]
4	179.7, C		
5	114.3, CH	5.99 d (2.2)	C-4w, C-6, C-7, C-3
6	168.5, C		
7	40.1, CH$_2$	3.69 s	C-5, C-6, C-8, C-9/C-13
8	138.9, C		
9	108.3, CH	6.30 d (2.1)	C-7, C-10, C-11, C-13
10	159.7, C		
11	102.2, CH	6.28 t (2.1)	C-9/C-13, C-10/C-12
12	159.7, C		
13	108.3, CH	6.30 d (2.1)	C-7, C-9, C-11, C-12
14	19.6, CH$_3$	2.22 d (0.5)	C-2, C-3

[a]w: weak correlation.

δ_C 108.3 (CH-9/CH-13), 102.2 (CH-11), 114.3 (CH-3/CH-5), 40.1 (CH$_2$-7), and 19.6 (CH$_3$-14) and six quaternary carbon signals that were distinguished by DEPT-135 and HSQC experiments. The chemically equivalent aromatic protons (H-9/H-13) showed meta coupling to the third aromatic proton (H-11) and long-range correlation to H$_2$-7 in the COSY spectrum, thus indicating the presence of a symmetrically trisubstituted benzene ring. The HMBC correlations from H$_2$-7 to CH-9/CH-13 and C-8 (δ_C 138.9) and from H-9/H-13 to C-10/C-12 (δ_C 159.7), CH-11, and CH$_2$-7 further suggested a 10,12-dihydroxy-8-methylenebenzene ring being present in 1. In the COSY spectrum, additional long-range correlations via four bonds were observed between H-3/H-5 and H-3/H$_3$-14. This was corroborated by the HMBC spectrum, in which the correlations from H$_3$-14 to C-3, from H-3 to C-14 and C-5, and

from H-5 to C-3 were discerned. Moreover, the HMBC correlations (Table 2) from H$_3$-14 to C-2 (δ_C 166.5), from H-3 to C-2 and the carbonyl (C-4, δ_C 179.7), from H-5 to C-4, C-6 (δ_C 168.5), and C-7, and from H$_2$-7 to C-6 and C-5 allowed the assignment of a 2-methyl and 6-methylene substituted 4-pyranone ring. Therefore, compound 1 is identified as 2-(3,5-dihydroxybenzyl)-6-methyl-4H-pyran-4-one, which is structurally related to the fungal metabolites macrocarpons A and B, which are produced by the fungi of the family Xylariaceae.[18,19] The trivial name macrocarpon C is given to this new compound.

Compound 3 exhibited a pseudomolecular peak at m/z 195.06067 [M + H]⁺ in the HRESIMS spectrum, corresponding to the molecular formula $C_9H_9NO_4$. The ¹H NMR spectrum of 3 revealed four aromatic proton signals at δ_H 7.91 (1H, dd, H-6), 7.36 (1H, ddd, H-4), 6.62 (1H, ddd, H-5), and 6.61 (1H, dd, H-3), which were assignable to an ABCD spin system and showed a singlet integrated as 2H at δ_H 4.01 (H$_2$-8) for a methylene group. The ¹³C NMR and DEPT spectra (Table 3) revealed the presence of one aromatic ring

Table 3. ¹H and ¹³C NMR Data and HMBC Correlations of 3 (CD$_3$OD)

position	δ_C, type	δ_H mult. (J in Hz)	HMBC correlations (H→C)
1	112.3, C		
2	151.9, C		
3	112.5, CH	6.61 dd (8.7, 1.0)	C-5, C-1
4	135.8, CH	7.36 ddd (8.7, 7.2, 1.6)	C-6, C-2
5	116.4, CH	6.62 ddd (7.8, 7.2, 1.0)	C-3, C-1
6	133.4, CH	7.91 dd (7.8, 1.6)	C-7, C-2, C-4
7	171.9, C		
8	45.5, CH$_2$	4.01 s	C-2, C-9
9	174.1, C		

(δ_C 151.9, 135.8, 133.4, 116.4, 112.5, 112.3), one methylene (δ_C 45.5, C-8), and two carbonyl groups (δ_C 174.1, 171.9). One carbonyl (δ_C 171.9, C-7) was located at C-1 of the benzene ring, since H-6 (δ_H 7.91, dd) showed an HMBC correlation with the former. The methylene group (H$_2$-8) showed HMBC correlations only to C-2 (δ_C 151.9) and the second carbonyl (δ_C 174.1, C-9), while no correlation from H-3 to the methylene group (C-8) was observed. Thus, C-2 and C-8 were connected through a nitrogen atom, and two carboxylic groups were present at C-7 and C-9 to complete the structure of 3. Compound 3 was thus identified as N-(carboxymethyl)-anthranilic acid. This compound was first introduced as a synthetic compound in 1890 by Heumann as a starting material for indigo synthesis.[20] Since then, it has been used in many organic syntheses as a starting material.[20–22] However, to the best of our knowledge, this is the first report of the isolation of 3 from nature. Biosynthetically, compound 3 is an anthranilic acid derivative, which could originate from condensation of anthranilic acid and acetyl Co-A. Anthranilic acid is a key building block of the peptides talaromins A and B produced by the fungus Talaromyces wortmannii[23] and of several alkaloids derived from fungi, such as fumiquinazoline A,[24] chaetominine,[25] asperlicin,[26] and cottoquinazoline A.[27] On the other hand, anthranilic acid was also reported as a natural product produced by Streptomyces sp.,[28,29] Paenibacillus polymyxa,[30] and other bacteria.[31] Hence, the producer of compound 3 during cocultivation of F. tricinctum and B. subtilis is still unclear.

Compounds 2 and 4 were likewise lacking either in axenic F. tricinctum or in B. subtilis. Compound 2 was identified as (−)-citreoisocoumarin by comparison of the spectroscopic data with the literature.[32,33] The NMR, MS, and UV spectra of compound 4 matched well with those reported for citreoisocoumarinol;[33] however, their optical rotation values were opposite ($[\alpha]_D$ −20.1 (4); +20.2 (lit.)[33]); thus 4 was determined as (−)-citreoisocoumarinol, which is a new natural product. Recently, the production of 2 was reported from F. graminearum, F. tricinctum, and other Fusarium species as presumably from a redirected pathway of bikaverin due to nitrogen starvation of the fungal cultures.[34] Hence, the production of 2 induced by coculture of F. tricinctum and B. subtilis might be a result of the competition for nutrients between fungus and bacterium.

The induced accumulation of fungal metabolites as reported in this study was found to correlate with the time of preincubation of the solid rice medium with B. subtilis prior to inoculation with F. tricinctum. When the fungus and the bacterium were inoculated on rice medium at the same day, no, or only a minor, effect on the accumulation of the fungal products was observed (data not shown). The strongest induction of fungal metabolites was detected following a six- or eight-day preincubation of the rice medium with B. subtilis, whereas this effect declined again after 10 days of preincubation (data not shown). For the experiments reported in this paper, a six-day preincubation of B. subtilis prior to inoculation with F. tricinctum was chosen. Under these experimental conditions the fungal growth as observed by visual inspection of the culture flasks was initially slowed down compared to axenic controls of F. tricinctum, but recovered after six to eight days of coculture, indicating an inhibitory effect caused by the presence of B. subtilis in the medium.

In a second set of experiments F. tricinctum was cocultured with Streptomyces lividans, a bacterial species that occurs in soil[35−37] but has also been reported as an endophyte of plants.[38] Also here, a clear induction of the accumulation of enniatins B (6), B1 (7), and A1 (8) as well as of fusaristatin A (9) was detected when compared to axenic controls of F. tricinctum (Figure 1). Compounds 1−4, however, were lacking. Instead, other compounds that remain to be identified were detected in extracts of F. tricinctum cocultured with S. lividans that were absent in fungal and bacterial controls (Figure 1). This indicates that the response of F. tricinctum to B. subtilis may at least partly be specific.

All isolated compounds were tested for their antibacterial activities (Table 4) against a broad spectrum of important Gram-positive and Gram-negative nosocomial pathogens. Enniatins B1 (7) and A1 (8) were active against B. subtilis 168 trpC2, the bacterial strain used for cocultivation, with MICs of 16 and 8 μg/mL, respectively, while enniatin B (6) did not inhibit the growth of this strain up to the highest concentration tested (64 μg/mL). Interestingly, the biosyntheses of these active enniatins, as opposed to the inactive enniatin B, were particularly upregulated by Fusarium during coculturing, leading to the tempting speculation that the fungus might have initiated their production in order to suppress its competitor. Enniatin B1 (7) further displayed MIC values of 8, 4, and 8 μg/mL against S. aureus, S. pneumoniae, and E. faecalis, respectively, and enniatin A1 (8) showed slightly lower MICs of 4−8, 2−4, and 4 μg/mL, respectively. The constitutively present lateropyrone (5) also displayed potent inhibitory activity against B. subtilis as well as against S. aureus, S.

Table 4. Minimal Inhibitory Concentration (MIC) [μg/mL] of the Isolated Compounds[a]

tested organism	resistance phenotype[a]	1	2	3	4	5[b]	6	7	8	9	vancomycin	ciprofloxacin	tetracycline
B. subtilis 168 trpC2	susceptible	>64	>64	>64	>64	8	>64	16	8	>64	0.125	0.06	4
S. aureus ATCC 29213	susceptible	>64	>64	>64	>64	2−4[c]	>64	8	4−8[c]	>64	0.5	0.125−0.0625	0.25
S. aureus 25697	AMX[a,d] CHL[R], CLI[R], CIP[R], ERY[R], FOS[R], GEN[R], KAN[R], NIT[R], TET[R] (MRSA[e])	>64	>64	>64	>64	2−4[c]	>64	8	4−8[c]	>64	0.5	8	32
S. pneumoniae ATCC 49619	susceptible	>64	>64	>64	>64	4−8[c]	16	4	2−4[c]	>64	0.125	0.25	0.0625
E. faecalis UW 2689	CLA[R], ERY[R], MXF[R], TEL[R] (VRE[f])	>64	>64	>64	>64	8	>64	8	4	>64	64	16	64
E. coli ATCC 25922	susceptible	>64	>64	>64	>64	>64	>64	>64	>64	>64	>64	0.006	0.5
P. aeruginosa B 63230	CAZ[R], CIP[R], CPM[R], GEN[R], IMI[R], MER[R], PIP/TAZ[R]	>64	>64	>64	>64	>64	>64	>64	>64	>64	>64	16	32

[a]Antibiotic abbreviations and breakpoints for resistance were applied according to the CLSI guidelines[41] AMX, amoxicillin; CAZ, ceftazidime; CHL, chloramphenicol; CIP, ciprofloxacin; CLA, clarithromycin; CLI, clindamycin; CPM, cefepime; ERY, erythromycin; FOS, fosfomycin; GEN, gentamycin; KAN, kanamycin; MER, meropenem; MXF, moxifloxacin; NIT, nitrofurantoin; PIP/TAZ, piperacillin/tazobactam; TEL, telithromycin; TET, tetracycline. [b]Compound 5 is hardly soluble in aqueous solvents, methanol, ethanol, and the used dimethyl sulfoxide; variations in outcome may be caused by this low solubility, and antimicrobial activity of 5 may be even higher if compound solubility can be improved. [c]The experiment was repeated six times and showed variations in the depicted ranges. [d]R: resistant. [e]MRSA: methicillin-resistant S. aureus. [f]VRE: vancomycin-resistant enterococci.

pneumoniae, and E. faecalis with MICs of 8, 2−4, 4−8, and 8 μg/mL, respectively. All other metabolites isolated in this study were inactive against all of the strains, tested up to a concentration of 64 μg/mL. A further notable aspect is that all antibacterial compounds described in this study retained full activity against multidrug-resistant staphylococcal and enterococcal clinical isolates (for details on the resistance phenotypes, see the footnote of Table 4). Lateropyrone (5) had been previously reported to inhibit the growth of S. aureus, although no quantitative value had been determined.[16] Regarding fusaristatin A (9), a previous publication had found this compound inactive against S. aureus and Pseudomonas aeruginosa at a concentration of 100 μg/mL,[15] which is in accordance with our results. For enniatin B (6), antibacterial activity was previously reported against Mycobacterium tuberculosis,[39] Clostridium perfringens CECT 4647, Salmonella enterica CECT 554, and S. aureus CECT 976.[40]

In conclusion, cocultivating of F. tricinctum and B. subtilis caused a strong enhancement of constitutively present fungal products and also stimulated the accumulation of several compounds that were not detected in either fungal or bacterial axenic controls. Most of these latter compounds (if not all) originate from F. tricinctum based on structural analogies with other known fungal products. Several fungal products identified in this study such as 5, 7, and 8 show antibacterial activity against several Gram-positive bacteria including MRSA. The induced production of compounds 1−4 was not duplicated when F. tricinctum was cocultivated with S. lividans. The latter treatment led rather to an accumulation of other, as yet unknown, compounds that were lacking in cocultures of F. tricinctum and B. subtilis. The fact that cocultivation in the present study enhanced the production of bioactive fungal products that inhibited the growth of the bacterial competitor demonstrates the general value of such cocultivating experiments and encourages further studies, even those including cocultivations with pathogenic bacteria.

■ EXPERIMENTAL SECTION

General Experimental Procedures. Optical rotations were determined on a Perkin-Elmer-241 MC polarimeter. 1D and 2D NMR spectra were recorded on an Avance DMX 600 NMR spectrometer. Chemical shifts were referenced to the residual solvent peak at δ_H 3.31 (CD$_3$OD) and 2.05 (acetone-d_6) for ^{1}H and δ_C 49.15 (CD$_3$OD) and 29.92 (acetone-d_6) for ^{13}C, respectively. Mass spectra were measured with a LCMS HP1100 Agilent Finnigan LCQ Deca XP Thermoquest, and high-resolution electrospray ionization mass spectroscopy (HRESIMS) were recorded with a UHR-TOF maXis 4G (Bruker Daltonics, Bremen) mass spectrometer. HPLC analysis was performed with a Dionex P580 system coupled to a photodiode array detector (UVD340S); routine detection was at 235, 254, 280, and 340 nm. The separation column (125 × 4 mm) was prefilled with Eurosphere-10 C18 (Knauer, Germany) and the following gradient was used (MeOH−H$_2$O (containing 0.1% HCOOH)): 0−5 min (10% MeOH); 5−35 min (10−100% MeOH); 35−45 min (100% MeOH). Semipreparative HPLC was performed using a Merck Hitachi HPLC System (UV detector L-7400; pump L-7100; Eurosphere-100 C$_{18}$, 300 × 8 mm, Knauer, Germany). Column chromatography was performed on Silica gel 60 M (230−400 mesh ASTM, Macherey-Nagel GmbH & Co. KG, Dueren, Germany) and Sephadex LH-20 (Sigma). TLC was carried out on precoated silica gel plates (silica gel 60 F-254, Merck KGaA, Darmstadt, Germany) for monitoring of fractions by using EtOAc−MeOH−H$_2$O (30:5:4) and CH$_2$Cl$_2$−MeOH (9:1) as solvent systems. Detection was at 254 and 366 nm or by spraying the plates with anisaldehyde reagent. Bacterial growth was monitored by measuring the OD$_{600}$ in a Tecan microtiter plate reader (Infinite M200, Tecan).

Microbial Material. The endophytic fungus was isolated from fresh, healthy rhizomes of Aristolochia paucinervis collected in January 2006 from the mountains of Beni-Mellal, Morocco, as previously described.[10] The bacterial strain panel included antibiotic-susceptible strains: a standard laboratory strain, Bacillus subtilis 168 trpC2;[41] the quality control strains of the Clinical Laboratory Standard Institute (CLSI), Staphylococcus aureus ATCC 29213, Streptococcus pneumoniae ATCC 49619, and Escherichia coli ATCC 25922;[43] and the following (multi)drug-resistant clinical isolates: Staphylococcus aureus 25697 (AiCuris, Wuppertal, Germany), Enterococcus faecalis UW 2689 (Wolfgang Witte, Robert Koch Institute, Wernigerode, Germany), and Pseudomonas aeruginosa B 63230.[42]

Cocultivation Experiment of F. tricinctum with B. subtilis 168 trpC2. Growth of fungus and bacteria in coculture for isolation and identification of metabolites was carried out in Erlenmeyer flasks (1 L). The fungal and bacterial strains were cultivated on solid rice media. Twenty-four Erlenmeyer flasks (eight flasks for F. tricinctum alone, eight for coculture of F. tricinctum and B. subtilis, and eight for B. subtilis alone) containing 60 mL of distilled water and 50 g of commercially available milk rice (Milch-Reis, ORYZA) each were autoclaved before inoculating the fungus and the bacterium.

B. subtilis was grown in lysogeny broth (LB). An overnight culture of B. subtilis was used to inoculate prewarmed LB medium (1:20), which was then incubated at 37 °C with shaking at 200 rpm to mid exponential growth phase (optical density at 600 nm (OD$_{600}$) of 0.2−0.4). A 10 mL amount of the bacterial culture was added to the rice medium, which was further incubated for 6 days at 37 °C. After this preincubation, F. tricinctum grown on malt agar (5 pieces, 1 cm × 1 cm) was added to the rice medium containing bacteria (after 6 days incubation) under sterile conditions. Fungal and bacterial controls were grown axenically on rice medium. Coculture and axenic cultures of F. tricinctum or B. subtilis were kept under static conditions at 23 °C until they reached their stationary phase of growth (2 weeks for controls of F. tricinctum and 3 weeks for cocultures). Then 300 mL of EtOAc was added to the cultures to stop the growth of cells followed by shaking at 140 rpm for 8 h. The cultures were then left overnight and filtered on the following day using a Büchner funnel. The EtOAc was removed under vacuum. Each extract was then dissolved in 50 mL of MeOH, and 20 μL of this was then injected into the analytical HPLC column.

Extraction and Isolation. The crude extract of cocultures of F. tricinctum and B. subtilis (2 g) was subjected to vacuum liquid chromatography (VLC) using mixtures of n-hexane−EtOAc followed by CH$_2$Cl$_2$−MeOH as the eluting solvent. Fraction 7, eluted with CH$_2$Cl$_2$−MeOH (9:1), was further purified using Sephadex LH-20 followed by semipreparative HPLC with MeOH−H$_2$O to yield 1 (2 mg), 2 (1.5 mg), 4 (0.8 mg), and 9 (4 mg). Fraction 10, eluted with CH$_2$Cl$_2$−MeOH (3:7), and fraction 12 (eluted with MeOH containing 0.1% TFA) were further purified by semipreparative HPLC using MeOH−H$_2$O as the mobile phase to afford 3 (2.3 mg) and 5 (1.8 mg), respectively. Metabolites 6−8 were identified by comparison with authentic standards employing HPLC and LC-MS.

Macrocarpon C (1): white, amorphous powder; UV (λ_{max}, MeOH) log ε 203 (3.5), 251 (3.1) nm; ^{1}H NMR (600 MHz, acetone-d_6) and ^{13}C NMR (150 MHz, acetone-d_6) see Table 2; ESIMS m/z 233.2 [M + H]$^+$, 231.3 [M − H]$^-$; HRESIMS m/z 233.08074 [M + H]$^+$ (calcd for C$_{13}$H$_{13}$O$_4$ 233.08084).

N-(Carboxymethyl)anthranilic acid (3): yellow, amorphous powder; UV (λ_{max}, MeOH) log ε 218 (4.1), 255 (3.7), 348 (3.4) nm; ^{1}H NMR (600 MHz, CD$_3$OD) and ^{13}C NMR (150 MHz, CD$_3$OD) see Table 3; ESIMS m/z 196.0 [M + H]$^+$, 194.3 [M − H]$^-$; HRESIMS m/z 196.06067 [M + H]$^+$ (calcd for C$_9$H$_{10}$NO$_4$ 196.06043).

(−)-Citreoisocoumarinol (4): yellow, amorphous powder; [α]$^{22}_D$ −20.1 (c 0.09, MeOH); ESIMS m/z 281.0 [M + H]$^+$, 279.4 [M − H]$^-$; UV and NMR data matched well with the published data.[33]

Antibacterial Assay. MIC values were determined by the broth microdilution method according to CLSI guidelines.[43] For preparation of the inoculum, the direct colony suspension method was used with an inoculum of 5×10^5 colony forming units/mL after the last dilution step. Compounds were added from stock solution (10 mg/mL in DMSO), resulting in a final DMSO amount of 0.64% at the highest antibiotic concentration tested (64 μg/mL). Serial 2-fold dilutions of antibiotics were prepared with DMSO being diluted along with the compounds.

■ ASSOCIATED CONTENT

ⓈSupporting Information
^{1}H, ^{13}C, and MS spectra for compounds 1 and 3 are available free of charge via the Internet at http://pubs.acs.org.

■ AUTHOR INFORMATION

Corresponding Authors
*Phone: +49 211 81 14187. Fax: +49 211 81 11923. E-mail: laidaowan123@gmail.com.
*Phone: +49 211 81 14163. E-mail: proksch@uni-duesseldorf. de.

Notes
The authors declare no competing financial interest.

■ ACKNOWLEDGMENTS

This study was supported by grants of the BMBF to P.P., by the German Research Foundation (FOR854, BR 3783/1-2) to H.B.-O., and by the iGRASP$_{seed}$ grant from the Research Center Juelich (Germany) and the University of Duesseldorf (Germany) to D.T. A.R.B.O. thanks DAAD for a doctoral scholarship. Technical assistance of H. Goldbach-Gecke is gratefully acknowledged. In addition, we thank H.-G. Sahl (University of Bonn, Germany), W. Witte (Robert-Koch Institute, Wernigerode, Germany), and AiCuris GmbH & Co. KG (Wuppertal, Germany) for providing clinical isolates.

■ REFERENCES

(1) Strobel, G.; Daisy, B.; Castillo, U.; Harper, J. *J. Nat. Prod.* 2004, 67, 257−268.
(2) Aly, A. H.; Debbab, A.; Proksch, P. *Appl. Microbiol. Biotechnol.* 2011, 90, 1829−1845.
(3) Brakhage, A. A.; Schroeckh, V. *Fungal Genet. Biol.* 2011, 48, 15−22.
(4) Scherlach, K.; Hertweck, C. *Org. Biomol. Chem.* 2009, 7, 1753−1760.
(5) Oh, D.-C.; Kauffman, C. A.; Jensen, P. R.; Fenical, W. *J. Nat. Prod.* 2007, 70, 515−520.
(6) Schroeckh, V.; Scherlach, K.; Nutzmann, H.-W.; Shelest, E.; Schmidt-Heck, W.; Schuemann, J.; Martin, K.; Hertweck, C.; Brakhage, A. A. *Proc. Natl. Acad. Sci. U.S.A.* 2009, 106, 14558−14563.
(7) Nützmann, H.-W.; Reyes-Dominguez, Y.; Scherlach, K.; Schroeckh, V.; Horn, F.; Gacek, A.; Schumann, J.; Hertweck, C.; Strauss, J.; Brakhage, A. A. *Proc. Natl. Acad. Sci. U.S.A.* 2011, 108, 14282−14287.
(8) Oh, D.-C.; Jensen, P. R.; Kauffman, C. A.; Fenical, W. *Bioorg. Med. Chem.* 2005, 13, 5267−5273.
(9) Cueto, M.; Jensen, P. R.; Kauffman, C.; Fenical, W.; Lobkovsky, E.; Clardy, J. *J. Nat. Prod.* 2001, 64, 1444−1446.
(10) Waetjen, W.; Debbab, A.; Hohlfeld, A.; Chovolou, Y.; Kampkoetter, A.; Edrada, R. A.; Ebel, R.; Hakiki, A.; Mosaddak, M.; Totzke, F.; Kubbutat, M. H. G.; Proksch, P. *Mol. Nutr. Food Res.* 2009, 53, 431−440.
(11) Yang, N.-Y.; Jiang, S.; Shang, E.-X.; Tang, Y.-P.; Duan, J.-A. *J. Chem. Res.* 2012, 36, 647.
(12) Goryluk, A.; Rekosz-Burlaga, H.; Blaszczyk, M. *Pol. J. Microbiol.* 2009, 58, 355−361.
(13) Tiwari, R.; Kalra, A.; Darokar, M. P.; Chandra, M.; Aggarwal, N.; Singh, A. K.; Khanuja, S. P. S. *Curr. Microbiol.* 2010, 60, 167−171.

(14) Wang, J.-P.; Lin, W.; Wray, V.; Lai, D.; Proksch, P. *Tetrahedron Lett.* 2013, 54, 2492−2495.
(15) Shiono, Y.; Tsuchinari, M.; Shimanuki, K.; Miyajima, T.; Murayama, T.; Koseki, T.; Laatsch, H.; Funakoshi, T.; Takanami, K.; Suzuki, K. *J. Antibiot.* 2007, 60, 309−316.
(16) Bushnell, G. W.; Li, Y. L.; Poulton, G. A. *Can. J. Chem.* 1984, 62, 2101−2106.
(17) Gorst-Allman, C. P.; Van Rooyen, P. H.; Wnuk, S.; Golinski, P.; Chelkowski, J. *S. Afr. J. Chem.* 1986, 39, 116−117.
(18) Laessoe, T.; Srikitikulchai, P.; Fournier, J.; Kopcke, B.; Stadler, M. *Fungal Biol.* 2010, 114, 481−489.
(19) Stadler, M.; Fournier, J.; Quang, D. N.; Akulov, A. Y. *Nat. Prod. Commun.* 2007, 2, 287−304.
(20) Wiklund, P.; Romero, I.; Bergman, J. *Org. Biomol. Chem.* 2003, 1, 3396−3403.
(21) Wang, Z. H.; Li, W. Y.; Li, F. L.; Zhang, L.; Hua, W. Y.; Cheng, J. C.; Yao, Q. Z. *Chin. Chem. Lett.* 2009, 20, 542−544.
(22) Lai, T. K.; Chatterjee, A.; Banerji, J.; Sarkar, D.; Chattopadhyay, N. *Helv. Chim. Acta* 2008, 91, 1975−1983.
(23) Bara, R.; Aly, A. H.; Wray, V.; Lin, W. H.; Proksch, P.; Debbab, A. *Tetrahedron Lett.* 2013, 54, 1686−1689.
(24) Numata, A.; Takahashi, C.; Matsushita, T.; Miyamoto, T.; Kawai, K.; Usami, Y.; Matsumura, E.; Inoue, M.; Ohishi, H.; Shingu, T. *Tetrahedron Lett.* 1992, 33, 1621−1624.
(25) Jiao, R. H.; Xu, S.; Liu, J. Y.; Ge, H. M.; Ding, H.; Xu, C.; Zhu, H. L.; Tan, R. X. *Org. Lett.* 2006, 8, 5709−5712.
(26) Haynes, S. W.; Gao, X.; Tang, Y.; Walsh, C. T. *J. Am. Chem. Soc.* 2012, 134, 17444−17447.
(27) Fremlin, L. J.; Piggott, A. M.; Lacey, E.; Capon, R. J. *Nat. Prod.* 2009, 72, 666−670.
(28) Abdelfattah, M. S.; Toume, K.; Arai, M. A.; Masu, H.; Ishibashi, M. *Tetrahedron Lett.* 2012, 53, 3346−3348.
(29) Yang, S.-W.; Cordell, G. A. *J. Nat. Prod.* 1997, 60, 44−48.
(30) Lebuhn, M.; Heulin, T.; Hartmann, A. *FEMS Microbiol. Ecol.* 1997, 22, 325−334.
(31) Maskey, R. P.; Asolkar, P. N.; Kapaun, E.; Wagner-Dobler, I.; Laatsch, H. *J. Antibiot.* 2002, 55, 643−649.
(32) Watanabe, A.; Ono, Y.; Fujii, I.; Sankawa, U.; Mayorga, M. E.; Timberlake, W. E.; Ebizuka, Y. *Tetrahedron Lett.* 1998, 39, 7733−7736.
(33) Lai, S.; Shizuri, Y.; Yamamura, S.; Kawai, K.; Furukawa, H. *Heterocycles* 1991, 32, 297−305.
(34) Soerensen, J. L.; Nielsen, K. F.; Sondergaard, T. E. *Fungal Genet. Biol.* 2012, 49, 613−618.
(35) Meschke, H.; Walter, S.; Schrempf, H. *Environ. Microbiol.* 2012, 14, 940−952.
(36) Adegboye, M. F.; Babalola, O. O.; Ngoma, L.; Okoh, A. I. *J. Pure Appl. Microbiol.* 2012, 6, 1001−1010.
(37) Mellado, R. P. *World J. Microbiol. Biotechnol.* 2011, 27, 2231−2237.
(38) Gulshan, M.; Zhao, G.-Y.; Dong, Z.-F.; Ghopur, M. *Shengwu Jishu* 2009, 19, 43−46.
(39) Nilanonta, C.; Isaka, M.; Chanphen, R.; Thong-orn, N.; Tanticharoen, M.; Thebtaranonth, Y. *Tetrahedron* 2003, 59, 1015−1020.
(40) Meca, G.; Sospedra, I.; Valero, M. A.; Mannes, J.; Font, G.; Ruiz, M. J. *Toxicol. Mech. Methods* 2011, 21, 503−512.
(41) Burkholder, P. R.; Giles, N. H. *Am. J. Bot.* 1947, 34, 345−348.
(42) Henrichfreise, B.; Wiegand, I.; Sherwood, K. J.; Wiedemann, B. *Antimicrob. Agents Chemother.* 2005, 49, 1668−1669.
(43) CLSI. *Methods for Dilution Antimicrobial Susceptibility Tests for Bacteria That Grow Aerobically*; Approved Standard Ninth ed.; CLSI document M07-A9; Clinical and Laboratory Standards Institute: Wayne, PA, 2012.

Supporting Information

Inducing Secondary Metabolite Production of the Endophytic Fungus *Fusarium tricinctum* through Co-culture with *Bacillus subtilis*

Antonius R.B. Ola [†,‡], Dhana Thomy[†], Daowan Lai[†,*], Heike Brötz-Oesterhelt[†], Peter Proksch[†,*]

[†] Institut für Pharmazeutische Biologie und Biotechnologie, Heinrich-Heine-Universität Düsseldorf, Universitätsstrasse 1, Geb. 26.23, 40225 Düsseldorf, Germany

[‡] Department of Chemistry, Faculty of Science and Engineering, Nusa Cendana University, Jalan Adisucipto Penfui, 85001 Kupang, Indonesia.

* Corresponding authors.

Tel.: +49 211 81 14187 (D.L.), +49 211 81 14163 (P.P.); Fax: +49 211 81 11923.

E-mail addresses: laidaowan123@gmail.com (D.L.), proksch@uni-duesseldorf.de (P.P.).

Table of Contents

Fig. S1-1 ^{1}H NMR spectrum of **1** in Acetone-d_6 (600 MHz) 3

Fig. S1-2 ^{13}C NMR spectrum of **1** in Acetone-d_6 (150 MHz) 3

Fig. S1-3 HRESIMS spectrum of **1** 4

Fig. S2-1 ^{1}H NMR spectrum of **3** in CD$_3$OD (600MHz) 5

Fig. S2-2 ^{13}C NMR spectrum of **3** in CD$_3$OD (150 MHz) 6

Fig. S2-3 DEPT-135 spectrum of **3** in CD$_3$OD (150 MHz) 7

Fig. S2-4 HRESIMS spectrum of **3** 8

Coculture

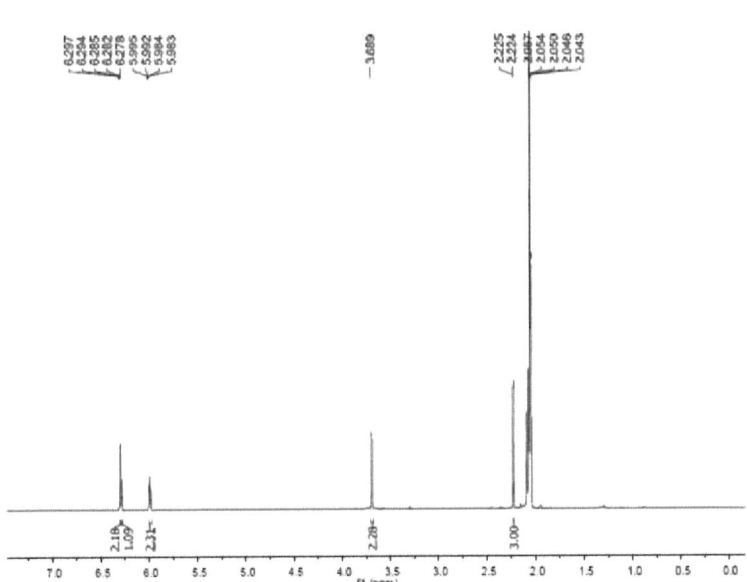

Fig. S1-1 ^{1}H NMR spectrum of **1** in Acetone-d_6 (600 MHz)

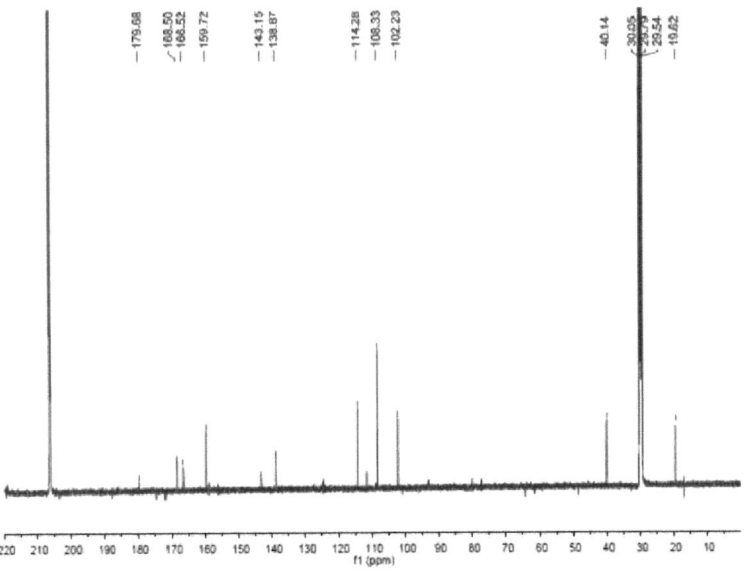

Fig. S1-2 ^{13}C NMR spectrum of **1** in Acetone-d_6 (150 MHz)

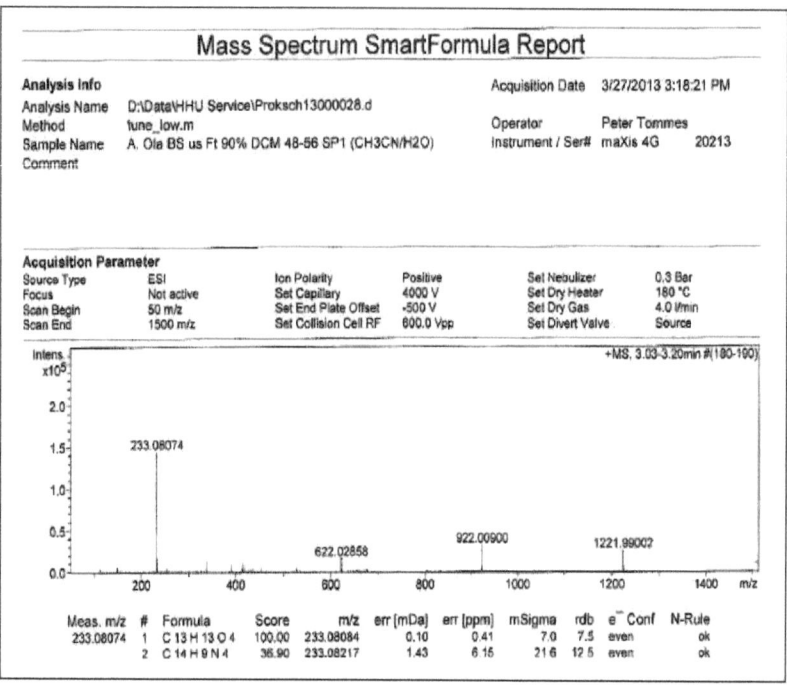

Fig. S1-3 HRESIMS spectrum of **1**

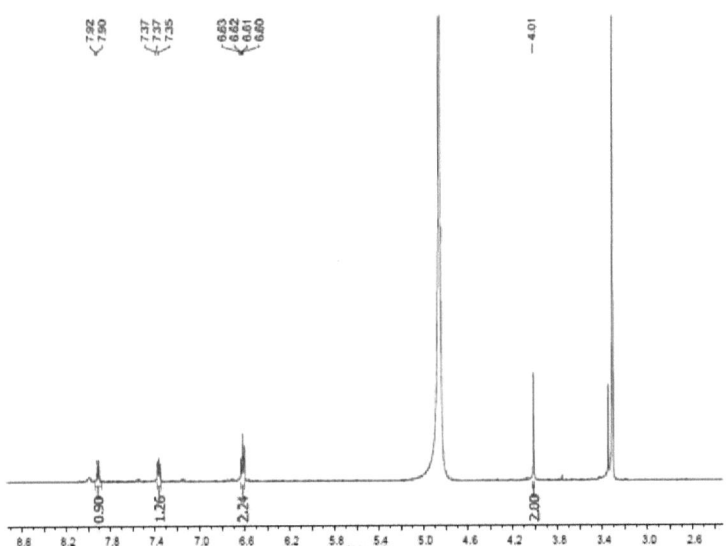

Fig. S2-1 ¹H NMR spectrum of **3** in CD$_3$OD (600MHz)

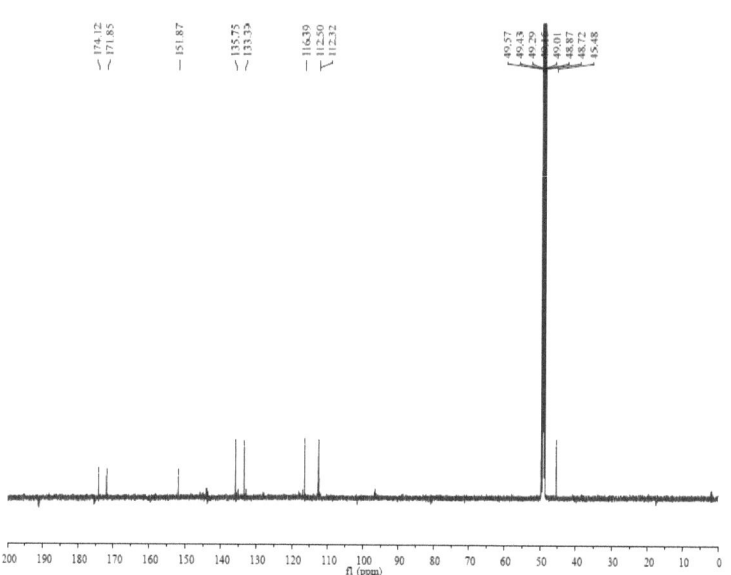

Fig. S2-2 ^{13}C NMR spectrum of **3** in CD$_3$OD (150 MHz)

Fig. S2-3 DEPT-135 spectrum of 3 in CD$_3$OD (150 MHz)

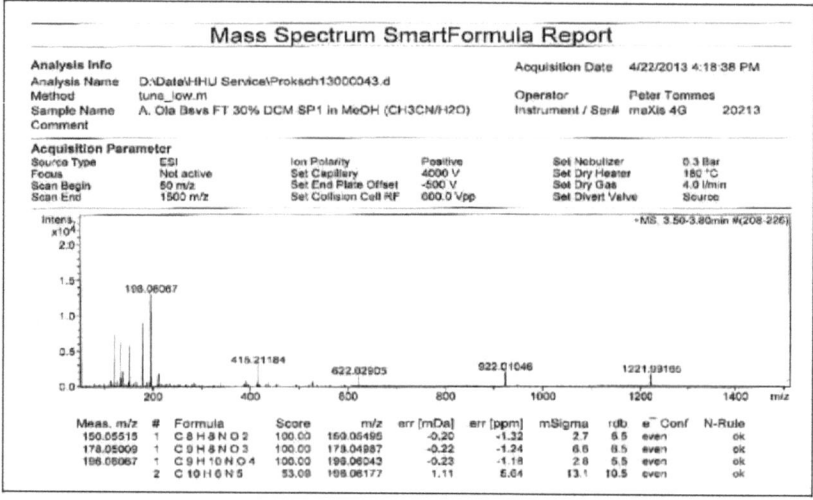

Fig. S2-4 HRESIMS spectrum of 3

Chapter 3

Absolute Configuration and Antibiotic Activity of Neosartorin from the Endophytic Fungus *Aspergillus fumigatiaffinis*

In Press, Accepted Manuscript, Available online 26 December 2013 in "Tetrahedron Letters"
Impact Factor: 2.397,
The overall contribution to the paper: 75 % of the first author. The first author involved to all laboratory works as well as the manuscript preparation.

Absolute Configuration and Antibiotic Activity of Neosartorin from the Endophytic Fungus *Aspergillus fumigatiaffinis*

Antonius R.B. Ola[a,d], Abdessamad Debbab[a], Amal H. Aly[a], Attila Mandi[c], Ilka Zerfass[a], Alexandra Hamacher[b], Matthias U. Kassack[b], Heike Brötz-Oesterhelt[a], Tibor Kurtan[c], Peter Proksch[a,*]

[a] Institute of Pharmaceutical Biology and Biotechnology, Heinrich Heine University Duesseldorf, Universitaetsstrasse 1, Geb. 26.23, 40225 Duesseldorf, Germany.

[b] Department of Organic Chemistry, University of Debrecen, POB 20, 4010 Debrecen, Hungary.

[c] Institute of Pharmaceutical and Medicinal Chemistry, Heinrich Heine University Duesseldorf, Universitaetsstrasse 1, Geb. 26.23, 40225 Duesseldorf, Germany.

[d] Department of Chemistry, Faculty of Science and Engineering, Nusa Cendana University, Jalan Adisucipto Penfui, 85001 Kupang, Indonesia.

---

[*] Corresponding author. Tel.: +49-211-81-14163; fax: +49-211-81-11923; e-mail: proksch@uni-duesseldorf.de

Abstract

Neosartorin (**1**) was isolated from the endophytic fungus *Aspergillus fumigatiaffinis*. The absolute configuration of **1**, including both axial and central chirality elements, was established as (a*R*,5*S*,10*R*,5'*S*,6'*S*,10'*R*) for the first time on the basis of its electronic circular dichroism (ECD) spectra aided with TDDFT-ECD calculations. Neosartorin (**1**) exhibited substantial antibacterial activity against a broad spectrum of Gram-positive bacterial species including staphylococci, streptococci, enterococci and *Bacillus subtilis* with minimal inhibitory concentrations in the range of 4 to 32 μg/mL. When the toxicity of **1** against eukaryotic cells was measured using a panel of different cancer cell lines such as HELA and BALB/3T3, the average IC_{50} values exceeded 32 μg/mL.

Keywords *Aspergillus fumigatiaffinis*, Neosartorin, Absolute configuration, Antibacterial activity.

Introduction

Bacterial resistance to clinically used antibiotics is spreading and therapy failures due to insufficient antibiotic potency occur with increasing frequency. Among Gram-positive bacteria methicillin-resistant *Staphylococcus aureus* (MRSA) is an important example of a multidrug-resistant pathogen that is common in hospitals as well as within the community and is resistant even against antibiotics that were launched during the last decade (e.g. linezolid and daptomycin resistance has already developed).[1,2] Other important examples are *Streptococcus pneumoniae* with isolates resistant to three or more antibiotics being reported from many countries,[3] and *Enterococcus faecium*, where resistances to aminopenicillins, aminoglycosides and vancomycin cause considerable concern.[4] Among Gram-negative bacteria the resistance situation is even more serious.[5] Consequently, there is an urgent need to discover and develop new antibiotics with novel modes of action that are not affected by current resistance mechanisms.

Since the discovery of penicillin, the first β-lactam antibiotic introduced on the drug market, from *Penicillium notatum*, fungi provided several important drug leads for curing life threatening diseases such as cancer, atherosclerosis, rejection of transplanted organs and others. Recent research findings indicate that endophytic fungi that inhabit higher plants as commensals or symbionts are promising sources of antibacterial agents.[6-11] Several atropisomeric natural products such as the glycopeptide antibiotic vancomycin,[12] skyrin, the bisanthracene derivatives flavomannins A-D,[13] talaromannins A and B,[13] and alterporriol D[14] were reported to be active against a panel of pathogenic microorganisms including MRSA.

In the course of our search for antibacterial agents from endophytic fungi, we investigated the fungal strain *Aspergillus fumigatiaffinis* isolated from *Tribulus terrestris* (Zygophyllaceae) collected in Uzbekistan. In the present work, the atropisomeric neosartorin (**1**), an asymmetric biaryl derivative having both central and axial chirality elements, was isolated as the active ingredient of the antibiotic extract obtained from *A. fumigatiaffinis*. Neosartorin (**1**) had been reported first as a yellow pigment produced by the soil mould *Neosartoria fischeri*,[15] and three years later the relative configuration of its two xanthene units were determined by measuring long-range interunit NOE contacts.[16] Interestingly, although the restricted rotation along the C-2-C-4' biaryl axis was identified, the possibility of atropisomers had not been discussed and absolute configuration of the central and axial chirality elements had not been deduced. In atropisomers, axial chirality plays a decisive role with respect to the biological activity.[17] Hence, the determination of the absolute configuration of the biaryl axis of bioactive atropisomers is an important issue during structure determination. In this context, we report here for the first time the axial chirality of neosartorin (**1**) and the absolute configuration of its central chirality elements together with its antibacterial activity against antibiotic susceptible bacterial strains and against multiresistant clinical isolates.

Results and Discussion

The ethyl acetate extract of *A. fumigatiaffinis* was subjected to chromatographic separation using different stationary phases including silica gel and Sephadex LH-20 followed by purification with semi-preparative HPLC. Neosartorin (**1**) (Figure 1) was isolated as yellow amorphous powder. Its UV spectrum closely resembled that of the known secalonic acids D and G.[18] ESI- and EI-MS indicated a molecular weight of 680 g/mol and its molecular formula was determined as $C_{34}H_{32}O_{15}$ based on the prominent signal detected at *m/z* 681.18157 [M+H]$^+$ in the HRESI MS.

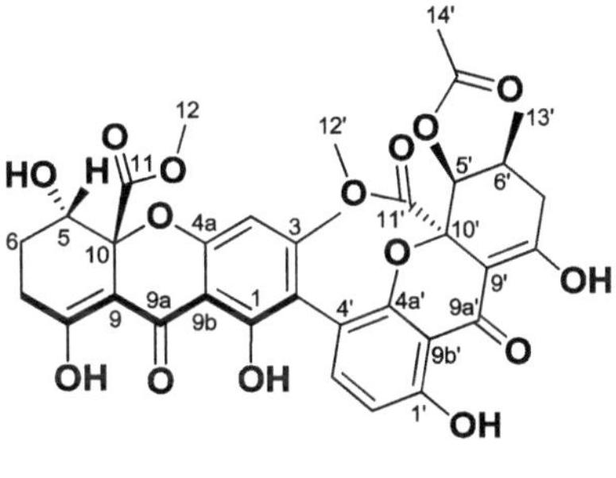

Figure 1. Structure of neosartorin (**1**).

Comparison of the MS and NMR data of **1** with those previously reported for neosartorin[15] suggested that they are identical. The structure of **1** was finally confirmed as neosartorin by careful analysis of COSY, HMBC and ROESY correlations. Inspection of the COSY and HMBC spectra suggested the presence of tetrahydroxanthone units as shown in Figure 2. The strong ^3J HMBC correlations from H-3' and H-4 to C-2 established the C-2—C-4' linkage of the two xanthone moieties. This connection was corroborated by ROESY correlations from both the methyl signal at C-3 and 1-OH to H-3', which also indicated that rings A and A' are not coplanar.

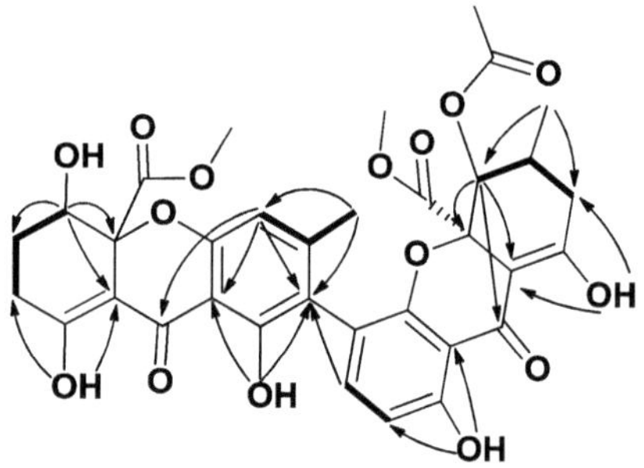

Figure 2. Important COSY (bold lines) and HMBC (arrows) correlations of neosartorin (**1**).

The relative configuration at C-5, C-10, C-5', C-6' and C-10' was deduced from the coupling constants extracted from the ^{1}H NMR spectrum and from ROESY experiments (Figure 3). CH$_3$-13' showed ROE correlations to both H-7'$_{ax}$ and H-7'$_{eq}$ which indicated its equatorial orientation and suggested a pseudoaxial configuration of H-6'. The latter correlated with the methoxcarbonyl group at C-10'. This supported the presence of H-6' and the methoxcarbonyl group on the same side of the six membered ring. H-5' showed ROE correlations to H-6' and CH$_3$-13' indicating its equatorial position, which was further confirmed by the small $^3J_{\text{H-6'—H-5'}}$ coupling (1.3 Hz). Thus, the acetoxy group at C-5' must have a pseudoaxial position.

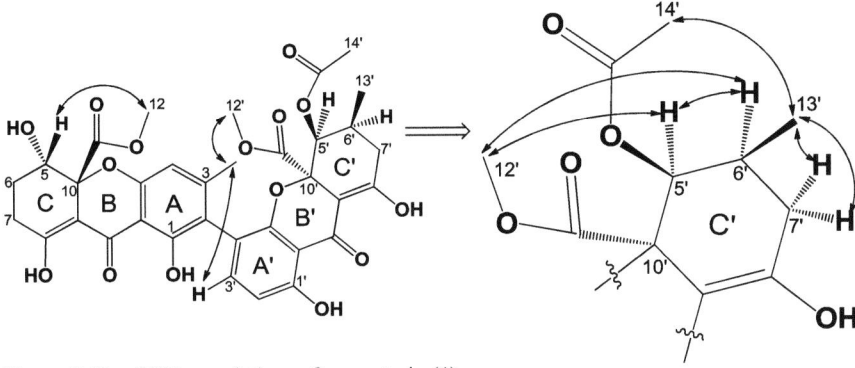

Figure 3. Key ROE correlations of neosartorin (**1**).

Similar to CH$_3$-13', H-5 correlated to H-6$_{ax}$ and H-6$_{eq}$ suggesting its equatorial position, which is confirmed by its small couplings ($^3J_{5-6ax}$= 4 and J_{5-6eq}= 1.8 Hz). In addition, the ROE correlation between H-5 and the methoxcarbonyl group at C-10 indicated their *cis* configuration. Thus, the relative configuration of **1** was established.

Electronic circular dichroism (ECD) has been found to be a powerful method to determine the axial chirality of atropisomeric biaryl natural products, since exciton coupled interaction between the two aryl units is sensitive to the sign and value of the biaryl torsional angle.[13,14,19] In combination with vibrational circular dichroism (VCD) absolute configuration of both axial and central chirality elements could be determined in the homodimeric cephalochromin A.[19] The same approach could determine only the absolute configuration of the axial chirality element in the symmetrical dimer flavomannin A, while its central chirality elements remained unresolved.[14] Thus, the configurational assignment of axial and central chirality elements in axially chiral biaryl natural products is considered a challenging task by spectroscopic methods.

The ECD spectrum of neosartorin (**1**) showed a characteristic intense negative exciton couplet centered around 334 nm, which derives from the interaction of the two 1-arylpropenone chromophores of the xanthene units (Figure 4A). Besides the couplet, negative Cotton effects (CEs) were observed at 235 and 200 nm and a positive one at 220 nm. In order to prove the axial chirality of neosartorin, a torsional angle scan was performed on (5S,10R,5'S,6'S,10'R)-**1**, since the (5S*,10R*,5'S*,6'S*,10'R*) relative configuration of the two xanthenes units was confirmed earlier by long-range NOE correlations.[16] Figure S1 shows the relative energies plotted in the function of the $\omega_{C-1,C-2,C-4',C-4a'}$ torsional angle. The estimated energy barriers (ca. 120 and 160 kJ/mol) indicated that the two atropodiastereomers cannot interconvert at room temperature; *i.e.* neosartorin has axial chirality. In accordance with Liptaj et al. results, we found that the (aR,5S,10R,5'S,6'S,10'R) atropodiastereomer has lower energy than the (aS,5S,10R,5'S,6'S,10'R) one.

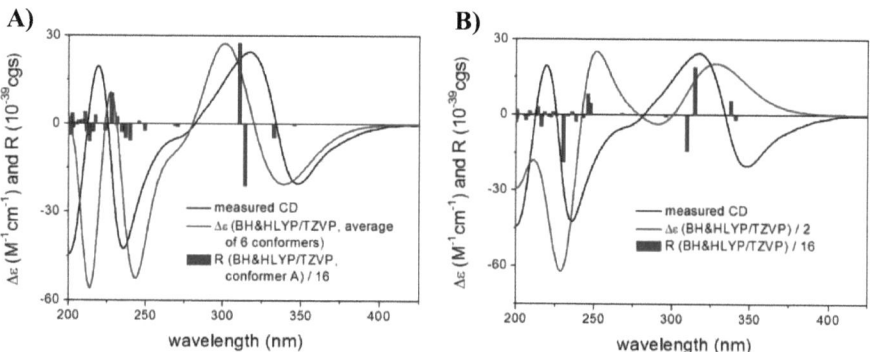

Figure 4. Experimental ECD spectrum of neosartorin (**1**) in acetonitrile compared with the Boltzmann-weighted BH&HLYP/TZVP spectrum calculated for (aR,5S,10R,5'S,6'S,10'R)-**1** (A) and wrong stereoisomer of **1** with (aS,5S,10R,5'S,6'S,10'R) absolute configuration (B) in vacuo. Computed ECD curves were shifted to the red with 30 nm.

A DFT reoptimization of the initial 15 MMFF conformers of (aR,5S,10R,5'S,6'S,10'R)-**1** resulted in 6 major conformers above 2% population (Figure S2), which differed mainly in the orientation of the CH$_3$-12 and CH$_3$-12' groups and hydroxyl protons. The $\omega_{C-1,C-2,C-4',C-4a'}$ torsional angle of the lowest-energy conformer of (aR,5S,10R,5'S,6'S,10'R)-**1** was −83.7°, while the higher-energy conformers had −82.18°, −94.5°, −71.6°, −80.8° and −77.0° values, respectively. The ECD spectra of the conformers were calculated with TZVP basis set and three different functionals (B3LYP, BH&HLYP, PBE0) resulting in consistent ECD curves. The Boltzmann-weighted ECD curves reproduced well the experimental ECD spectrum with BH&HLYP/TZVP giving the best agreement (Figures 4A, S3 and S4). Since the ECD spectrum of neosartorin is mostly governed by the axial chirality, the agreement of the computed and experimental ECD curves allowed determining the axial chirality as (aR). Thus the possible stereoisomers of neosartorin were restricted to the (aR,5S,10R,5'S,6'S,10'R) and (aR,5R,10S,5'R,6'R,10'S) diastereomers. For neosartorin, long-range interunit NOEs were observed for CH$_3$-14'/1-OH, CH$_3$-12'/3-CH$_3$, CH$_3$-12/CH$_3$-14' and CH$_3$-12/H-5', interatomic distances of which were found 2.68, 4.52, 3.41 and 4.16 Å, respectively, in the lowest-energy computed conformer of (aR,5S,10R,5'S,6'S,10'R) (Figure 5). It has to be noted that all the conformers contribute to the observed long-range NOEs and although for the sake of simplicity only were the interatomic distances of the lowest-energy conformers were discussed, some observed NOEs probably derives from minor conformers. For instance, the CH$_3$-12'/3-CH$_3$ NOE most likely derives from conformer B, in which 10'-methyloxycarbonyl group has a different orientation moving closer CH$_3$-12' and 3-CH$_3$ to each other. In contrast, CH$_3$-14'/1-OH, CH$_3$-12/CH$_3$-14' and CH$_3$-12/H-5' correlations are not feasible in computed conformers of the (aS,5S,10R,5'S,6'S,10'R) stereoisomer, since the 10- and 10'-methoxcarbonyl groups were located at opposite sides (Figure S5). This implies that these correlations are not possible in the enantiomeric (aR,5R,10S,5'R,6'R,10'S) stereoisomer either, and thus the absolute configuration of neosartorin (**1**) was determined as (a*R*,5*S*,10*R*,5'*S*,6'*S*,10'*R*). It is noteworthy that the atropodiastereomeric (a*S*,5*S*,10*R*,5'*S*,6'*S*,10'*R*) stereoisomer gave a near mirror image ECD curve of the experimental spectrum of neosartorin, but the agreement of the shape and transition maxima (Figure 4B) was much worse than that of the (a*R*,5*S*,10*R*,5'*S*,6'*S*,10'*R*) stereoisomer.

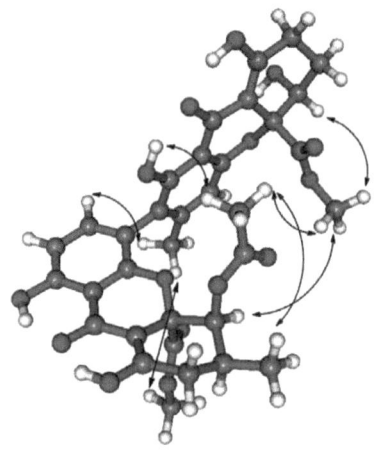

Figure 5. Key long range NOEs on the lowest-energy computed conformer (a*R*,5*S*,10*R*,5'*S*,6'*S*,10'*R*)-**1**.

Antibacterial Activity of Neosartorin

As part of our biological profiling neosartorin was evaluated for its antibacterial activity against a spectrum of Gram-positive and Gram-negative bacterial species, including antibiotic-susceptible reference strains as well as multi-resistant clinical isolates, and Minimal Inhibitory Concentration (MIC) values were determined by broth microdilution assays (Table 1). Neosartorin inhibited the growth of a broad spectrum of Gram-positive bacteria, while the tested Gram-negative species were not affected. MIC values from 4 to 8 µg/mL were obtained for S. aureus, B. subtilis and most streptococci and the growth of enterococci was inhibited at slightly higher concentrations (16 - 32 µg/mL). It is noteworthy that multiresistant clinical isolates were as susceptible as quality control strains, the latter being recommended by the Clinical Laboratory Standards Institute (CLSI)[20] for sensitive antibiotic testing.

Table 1. Minimal inhibitory concentrations [μg/mL] of **1**.

Tested organism	Resistance phenotype[a]	MIC
Gram-positive		
Staphylococcus aureus ATCC 29213	susceptible	8
Staphylococcus aureus Mu50	CAZ[R,b], CLA[R], CIP[R], CLI[R], DOX[R], ERY[R], MET[R], MXF[R], KAN[R], TEL[R], RIF[R] (MRSA[c], VISA[d])	8
Staphylococcus aureus 25697	AMX[R], CHL[R], CIP[R], CLI[R], ERY[R], FOS[R], GEN[R], KAN[R], NIT[R], TET[R] (MRSA)	8
Streptococcus pneumoniae ATCC 49619	susceptible	4
Streptococcus agalactiae 013761	DOX[I,e]	16
Streptococcus pyogenes 014327	DOX[I]	4
Enterococcus faecalis UW 2689	CLA[R], ERY[R], MXF[R], TEL[R], (VRE[f])	16
Enterococcus faecium 6011	CLA[R], ERY[R], TEL[R] (VRE)	32
Bacillus subtilis 168 trpC2	susceptible	4
Gram-negative		
Pseudomonas aeruginosa B 63230	CAZ[R], CIP[R], CPM[R], GEN[R], IMI[R], MER[R], PIP/TAZ[R]	>64
Escherichia coli ATCC 25922	susceptible	>64
Escherichia coli WT-3-1 MB2	CIP[R]	>64
Klebsiella pneumoniae ATCC 27799	DOX[R], KAN[R]	>64

Antibiotic neosartorin from *A. fumigatiaffinis*

[a]Antibiotic abbreviations and breakpoints for resistance were applied according to the CLSI guidelines (CLSI M7-A8, 2008): AMX (amoxicillin), CAZ (ceftazidime), CHL (chloramphenicol), CIP (ciprofloxacin), CLA (clarithromycin), CLI (clindamycin), CPM (cefepime), DOX (doxycycline), ERY (erythromycin), FOS (fosfomycin), GEN (gentamycin), KAN (kanamycin), LNZ (linezolid), MER (meropenem), MET (methicillin), MXF (moxifloxacin), NIT (nitrofurantoin), PIP/TAZ (piperacillin/tazobactam), RIF (rifampicin), TEL (telithromycin), TET (tetracycline). [b]R: Resistant. [c]MRSA: Methicillin-resistant *Staphylococcus aureus*. [d]VISA: Vancomycin-intermediate *Staphylococcus aureus*. [e]I: Intermediate. [f]VRE: Vancomycin-resistant enterococci.

Cytotoxic Activity of Neosartorin

When tested against A2780 sens/CisR and K 562 cancer cell lines at a concentration of 68 μg/mL, neosartorin (**1**) showed no activity. Neosartorin was further tested for putative cytotoxic activity against THP-1 human leukemic monocyte cells, HELA human cervix carcinoma cells and BALB/3T3 mouse embryonic fibroblast cells. The IC_{50} values for HELA and BALB/3T3 cells exceeded 32 μg/mL, the highest concentration tested. The THP-1 cell line, which is rather sensitive for growth inhibitory agents in general, showed an IC_{50} value of 12 μg/mL. Thus, the cytotoxic profile of neosartorin (**1**) indicated that it did not markedly interfere with growth and viability of the eukaryotic cells in the tested concentration range.

Acknowledgment

We thank Dr. Elena Kamilova (Institute of Genetics and Plant Experimental Biology, Academy of Sciences Republic of Uzbekistan, Kibrai District, P.O. Yukori-Yuz, 102151 Tashkent Region, Uzbekistan) for the supply and identification of plant material in this study. We also gratefully acknowledge Katharina Arenz (University of Duesseldorf, Germany) for performing MIC determinations for selected isolates and Peter Heisig (University of Hamburg, Germany), Hans-Georg Sahl (University of Bonn, Germany), Wolfgang Witte (Robert-Koch Institute, Wernigerode, Germany) and AiCuris GmbH & Co. KG (Wuppertal, Germany) for providing clinical isolates. A.R.B.O. wishes to thank the DAAD for the doctoral fellowship. This study was further supported by grants of the BMBF awarded to P.P. and A.D, and also by the grant of the state North Rhine-Westphalia and the European Union (European Regional Development Fund, Investing in your future) to I.Z. and H.B.-O. T.K. thanks the Hungarian National Research Foundation (OTKA K105871) for financial support and the National Information Infrastructure Development Institute (NIIFI 10038) for CPU time. A.M. also thanks for the Zoltán Magyary postdoctoral fellowship program. The computational stereochemical studies were realized in the frames of TÁMOP 4.2.4. A/2-11-1-2012-0001 National Excellence Program – Elaborating and operating an inland student and researcher personal support system convergence program. The project was subsidized by the European Union and co-financed by the European Social Fund.

Supplementary data

Supplementary data (figures of the torsional angle scan for estimating the rotational energy barrier around the C2-C4' bond of (aR,5S,10R,5'S,6'S,10'R)-**1**, of the low-energy conformers ($\geq$ 2%) of (aR,5S,10R,5'S,6'S,10'R)-**1**, and of the lowest-energy computed conformer of (aS,5S,10R,5'S,6'S,10'R)-**1**, as well as experimental section and compound characterization) associated with this article can be found in the online version, at http://XXXXXXXXXXXXXX

References

1. Stefani, S.; Chung, D. R.; Lindsay, J. A.; Friedrich, A. W.; Kearns, A. M.; Westh, H.; MacKenzie, F. M. *Int. J. Antimicrob. Agents* **2012**, *39*, 273-282.

2. Gould, I. M.; David, M. Z.; Esposito, S.; Garau, J.; Mazzei, T.;Peters, G. *Int. J. Antimicrob. Agents* **2012**, *39*, 96-104.

3. Reinert, R. R. *Clin. Microbiol. Infect.* **2009**, *15*, 7-11.

4. Arias, C. A.; Murray, B. E. *Nat. Rev. Microbiol.* **2012**, *10*, 266-278.

5. Livermore, D. M. *J. Antimicrob. Chemother.* **2009**, *64*, i29-i36.

6. Debbab, A.; Aly, A. H.; Lin, W.; Proksch, P. *J. Microb. Biotechnol.* **2010**, *3*, 544–563.

7. Aly, A. H; Debbab, A.; Kjer, J; Proksch, P. *Fungal Divers.* **2010**, *41*, 1-16.

8. Debbab, A.; Aly, A. H.; Proksch, P. *Fungal Divers.* **2011**, *49*, 1-12.

9. Aly, A. H; Debbab, A.; Proksch, P. *Appl. Microbiol. Biotechnol.* **2011**, *90*, 1829-1845.

10. Debbab, A.; Aly, A. H.; Proksch, P. *Fungal Divers.* **2012**, *57*, 45-83.

11. Debbab, A.; Aly, A. H.; Proksch, P. *Fungal Divers.* **2013**, 1-27.

12. Lloyd-Williams, P.; Giralt, E. *Chem. Soc. Rev.* **2001**, *30*, 145-157.

13. Bara, R.; Zerfass, I.; Aly, A. H.; Goldbach-Gecke, H.; Raghavan, V.; Sass, P.; Mandi, A.; Wray, V.; Polavarapu, P. L.; Pretsch, A.; Lin, W.; Kurtan, T.; Debbab, A.; Broetz-Oesterhelt, H.; Proksch, P. *J. Med. Chem.* **2013**, *56*, 3257-3272.

14. Debbab, A.; Aly, A. H.; Edrada-Ebel, R.; Wray, V.; Pretsch, P.; Pescitelli, G; Kurtan, T.; Proksch, P. *Eur. J. Org. Chem.* **2012**, *7*, 1351-1359.

15. Proksa, B.; Uhrin, D.; Liptaj, T.; Sturdikova, M. *Phytochemistry.* **1998**, *48*, 1161-1164.

16. Liptaj, T.; Pham, T. N.; Proksa, B.; Uhrin, D. *Chirality.* **2001**, *13*, 545-547.

17. LaPlante, S. R.; Fader, L. D.; Fandrick, K. R.; Fandrick, D. R.; Hucke, O.; Kemper, R.; Miller, S. P. F.; Edwards, P. *J. Med. Chem.* **2011**, *54*, 7005–7022.

18. Andersen, R.; Buechi, G.; Kobbe, B.; Demain, A. L. *J. Org. Chem.* **1977**, *42*, 352-353.

19. Polavarapu, P. L.; Jeirath, N.; Kurtan, T.; Pescitelli, G.; Krohn, K. *Chirality* **2009**, *21*, E202-E207.

20. CLSI. Methods for Dilution Antimicrobial Susceptibility Tests for Bacteria That Grow Aerobically; Approved Standart – Ninth Edition. CLSI document M07-A9. Wayne, PA: Clinical and Laboratory Standarts Institute; 2012.

Supporting Information

Absolute Configuration and Antibiotic Activity of Neosartorin from the Endophytic Fungus *Aspergillus fumigatiaffinis*

Antonius R.B. Ola[a,d], Abdessamad Debbab[a], Amal H. Aly[a], Attila Mandi[c], Ilka Zerfass[a], Alexandra Hamacher[b], Matthias U. Kassack[b], Heike Brötz-Oesterhelt[a], Tibor Kurtan[c], Peter Proksch[a,*]

[a] Institute of Pharmaceutical Biology and Biotechnology, Heinrich Heine University Duesseldorf, Universitaetsstrasse 1, Geb. 26.23, 40225 Duesseldorf, Germany.

[b] Department of Organic Chemistry, University of Debrecen, POB 20, 4010 Debrecen, Hungary.

[c] Institute of Pharmaceutical and Medicinal Chemistry, Heinrich Heine University Duesseldorf, Universitaetsstrasse 1, Geb. 26.23, 40225 Duesseldorf, Germany.

[d] Department of Chemistry, Faculty of Science and Engineering, Nusa Cendana University, Jalan Adisucipto Penfui, 85001 Kupang, Indonesia.

Table of Contents

Figure S1. Torsional angle scan for estimating the rotational energy barrier around the C2-C4' bond of (aR,5S,10R,5'S,6'S,10'R)-**1**.   S2

Figure S2. Low-energy conformers ($\geq$ 2%) of (aR,5S,10R,5'S,6'S,10'R)-**1**.   S3

Figure S3. Experimental ECD spectrum of neosartorin (**1**) in acetonitrile compared with the Boltzmann-weighted B3LYP/TZVP and PBE0/TZVP spectra calculated for (aR,5S,10R,5'S,6'S,10'R)-**1**.   S4

Figure S4. Experimental ECD spectrum of neosartorin (**1**) in acetonitrile compared with the Boltzmann-weighted B3LYP/TZVP and PBE0/TZVP spectra calculated for the wrong stereoisomer of **1** with (aS,5S,10R,5'S,6'S,10'R) absolute configuration.   S5

Figure S5. Lowest-energy computed conformer of (aS,5S,10R,5'S,6'S,10'R)-**1**.   S6

Experimental Section   S7

References   S11

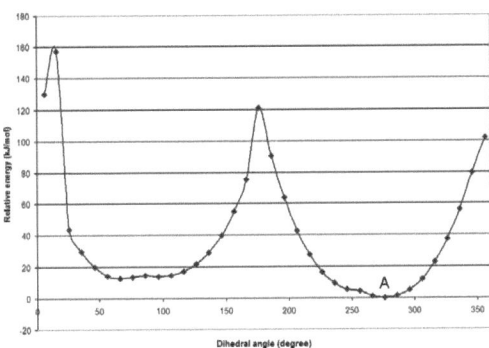

Figure S1. Torsional angle scan for estimating the rotational energy barrier around the C2-C4' bond ($\omega_{C-1,C-2,C-4',C-4a'}$ torsional angle) of (aR,5S,10R,5'S,6'S,10'R)-**1**. The scans were started from the lowest-energy in vacuo conformer (conformer A). Relative energy (kJ/mol) is plotted in the function of the $\omega_{C-1,C-2,C-4',C-4a'}$ torsional angle. The energy barriers for the inversion of axial chirality were found 120 and 160 kJ/mol.

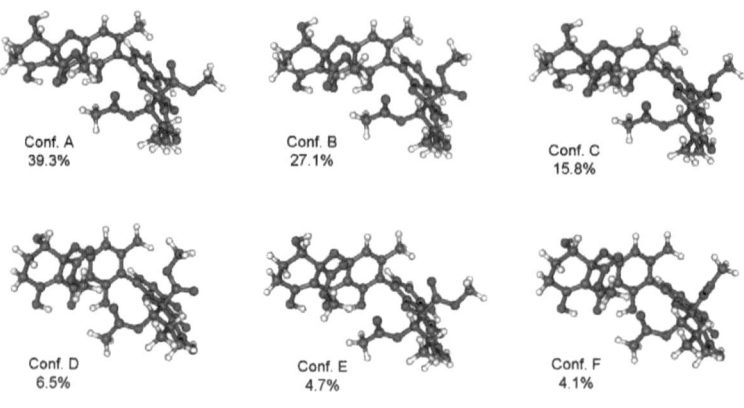

Figure S2. Low-energy conformers (≥ 2%) of (aR,5S,10R,5'S,6'S,10'R)-**1** calculated at B3LYP/6-31G(d) level of theory in vacuo.

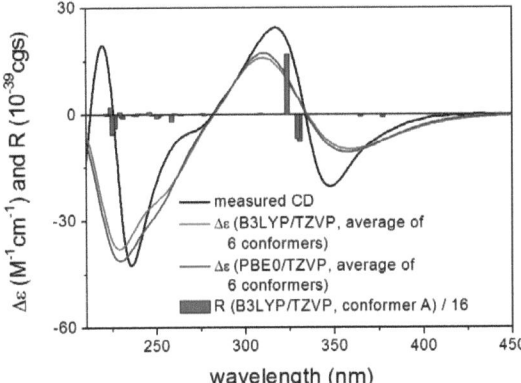

Figure S3. Experimental ECD spectrum of neosartorin (**1**) in acetonitrile compared with the Boltzmann-weighted B3LYP/TZVP (shifted with +8 nm) and PBE0/TZVP (shifted with +14 nm) spectra calculated for (a*R*,5*S*,10*R*,5'*S*,6'*S*,10'*R*)-**1**.

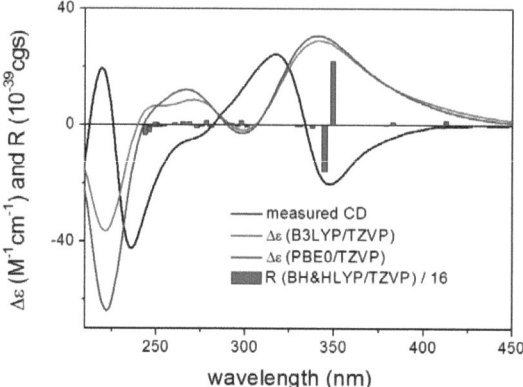

Figure S4. Experimental ECD spectrum of neosartorin (**1**) in acetonitrile compared with the Boltzmann-weighted B3LYP/TZVP (shifted with +8 nm) and PBE0/TZVP (shifted with +14 nm) spectra calculated for the wrong stereoisomer of **1** with (aS,5S,10R,5'S,6'S,10'R) absolute configuration.

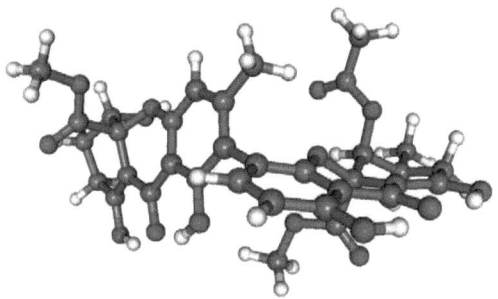

Figure S5. Lowest-energy computed conformer of (aS,5S,10R,5'S,6'S,10'R)-**1** in vacuo indicating that CH$_3$-14'/1-OH, CH$_3$-12/CH$_3$-14' and CH$_3$-12/H-5' long-range NOE correlations are not feasible in this stereoisomer.

Experimental Section

General Experimental Procedures

Optical rotations were determined on a Perkin-Elmer-241 MC polarimeter. 1D and 2D NMR spectra were recorded on a Bruker ARX 500 NMR spectrometer. Mass spectra were measured with a LCMS HP1100 Agilent Finnigan LCQ Deca XP Thermoquest and high-resolution electrospray ionization mass spectroscopy (HRESIMS) were recorded with an UHR-TOF maXis 4G (Bruker Daltonics, Bremen) mass spectrometer. HPLC analysis was performed with a Dionex P580 system coupled to a photodiode array detector (UVD340S); routine detection was at 235, 254, 280, and 340 nm. The separation column (125 × 4 mm) was prefilled with Eurosphere-10 C18 (Knauer, Germany), and the following gradient was used (MeOH, 0.1% HCOOH in H_2O): 0-5 min (10% MeOH); 5-35 min (10-100% MeOH); 35-45 min (100% MeOH). Semi-preparative HPLC was performed using a Merck Hitachi HPLC System (UV detector L-7400; Pump L-7100; Eurosphere-100 C18, 300 × 8 mm, Knauer, Germany). Column chromatography was performed on Silica gel 60 M (230−3400 mesh ASTM, Macherey-Nagel GmbH & Co. KG, Dueren, Germany) and Sephadex LH-20 (Sigma). TLC was carried out on precoated silica gel plates (silica gel 60 F-254, Merck KGaA, Darmstadt, Germany) for monitoring of fractions by using EtOAc/MeOH/H_2O (30:5:4) and CH_2Cl_2/MeOH (9:1) as solvent systems. Detection was at 254 and 366 nm or by spraying the plates with anisaldehyde reagent. HPLC/ECD analysis: HPLC separations were carried out with a Jasco HPLC system on a Chiralpak IC column (5 μm, 150×4.6 mm) using hexane/isopropanol 6:4 as eluent. The HPLC/ECD and HPLC/UV traces were recorded at 270 nm with a Jasco J-810 CD spectropolarimeter equipped with a 1 cm pathlength HPLC flow cell and the chromatogram was zeroed right after the start of recording, and hence relative absorbance was measured. The on-line CD and UV spectra (200–400 nm) were recorded simultaneously at the maxima of the UV peaks where the flow was stopped. ECD ellipticity values were not corrected for concentration. For an HPLC-ECD spectrum, three consecutive scans were recorded and averaged with 2 nm bandwidth, 1 s response, and standard sensitivity. The HPLC-ECD spectrum of the eluent recorded in the same way was used as background. The UV-absorption trace was recorded as high-tension voltage (HTV) and converted to absorbance. The concentration of the injected sample was set so that the HT value did not exceed 500 V in the HT channel down to 220 nm.

Fungal Material

Fresh, healthy leaves of *Tribulus terrestris* (Zygophyllaceae) were collected and identified by Dr. Elena Kamilova in Uzbekistan (2010). Leaves were rinsed twice with sterilized distilled water. Surface sterilization was achieved by immersing the leaves in 70% ethanol for 2 min (twice) followed by rinsing twice in sterilized distilled water. The leaves were then cut aseptically into small segments (approx.1 cm in length). The material was placed on a Petri dish (malt agar medium) containing chloramphenicol to suppress bacterial growth (medium composition: 15 g/L malt extract, 15 g/L agar, and 0.2 g/L chloramphenicol in distilled water, pH 7.4-7.8) and incubated at room temperature (22 °C). After several days, hyphae growing from the plant material were transferred to fresh plates with the same medium, incubated again for 10 days, and periodically checked for culture purity.

Identification of Fungal Cultures

Fungal cultures were identified according to a molecular biological protocol by DNA amplification and sequencing of the ITS region as described previously.[1] The sequence data have been submitted to GenBank, accession number HF545316. The fungal strain was identified as *Aspergillus fumigatiaffinis*. A voucher strain (strain designation EK 12.1.2) is kept in the Institute of Pharmaceutical Biology and Biotechnology, Duesseldorf, Germany.

Cultivation

Mass growth of fungus for isolation and identification of metabolites was carried out in Erlenmeyer flasks (1L). The fungal strain was cultivated on solid rice media. Two Erlenmeyer flasks containing 120 mL of distilled water and 100 g of commercially available rice each were autoclaved before inoculating the fungus. A small part of the medium from a Petri dish containing the purified fungus was transferred under sterile conditions to the rice medium. The fungal strain was grown on solid rice medium at room temperature (22 °C) for 28 days.

Extraction and Isolation

The crude extract of *A. fumigatiaffinis* (2 g) was subjected to vacuum liquid chromatography (VLC) using mixtures of n-hexane/EtOAc followed by CH_2Cl_2/MeOH as the eluting solvent. Fraction 7, eluted with CH_2Cl_2/MeOH (9:1), was further purified using Sephadex LH-20 followed by semi preparative HPLC with MeOH/H_2O to yield **1** (15 mg).

Neosartorin (1). Yellow amorphous powder; $[\alpha]^{22}_D$ -164.1 (*c*. 0.35, CHCl$_3$); UV λ_{max} (PDA) 211, 279 and 334 nm; ECD (CH$_3$CN, c = 1.6×10^{-4}) λ_{max} ($\Delta\varepsilon$): 374sh (−7.0), 347 (−20.5), 317 (24.8), 269sh (−5.2), 235 (−44.4), 220 (21.5), 200 (−45.9); ESIMS *m/z* 681.4 [M+H]$^+$;HRESIMS *m/z* 681.18157 [M+H]$^+$ (calcd for C$_{34}$H$_{33}$O$_{15}$ 681.18140).

Determination of Antibacterial Activity

MIC values were determined by the broth microdilution method according to CLSI guidelines.[2] For preparation of the inoculum the direct colony suspension method was used. The strain panel included antibiotic-susceptible CLSI quality control strains: *S. aureus* ATCC 29213, *Streptococcus pneumoniae* ATCC 49619, *E. coli* ATCC 25922, *Klebsiella pneumoniae* ATCC 27799,[2] a standard laboratory strain (*B. subtilis* 168 trpC2),[3] a high-level quinolone-resistant laboratory mutant (*E. coli* WT-3-1 MB2, Peter Heisig, University of Hamburg, Germany) and the following (multi)drug-resistant clinical isolates: *S. aureus* Mu50,[4] *S. aureus* 25697 (AiCuris, Wuppertal, Germany), *Streptococcus agalactiae* 013761 and *Streptococcus pyogenes* 014327 (Hans-Georg Sahl, University of Bonn, Germany), *Enterococcus faecalis* UW 2689 (Wolfgang Witte, Robert Koch Institute, Wernigerode, Germany), *Enterococcus faecium* 6011[5] and *P. aeruginosa* B 63230.[6]

Cytotoxicity Assay

THP-1 cells and BALB/3T3 cell were cultured as described.[7] HELA cells were grown in DMEM containing 4.5 g/l glucose and 4 mg/l pyridoxine hydrochloride (DMEM high glucose, GIBCO), supplemented with 10% foetal calf serum (v/v), 1% L–glutamine, 1 % penicillin-streptomycin (10,000 U/ml penicillin and 10 mg/ml streptomycin) (PAN Biotech GmbH) and 3.7 g/l NaHCO$_3$. The proliferative and metabolic capacity of the cell lines was measured as described[7] after an incubation of 72h in the presence of 0.016 to 32 µg/ml neosartorin and readout by the AlamarBlue assay as described.[7]

Computational Section

Mixed torsional/low mode conformational searches were carried out by means of the Macromodel 9.9.223[8] software using Merck Molecular Force Field (MMFF) with implicit solvent model for chloroform applying a 21 kJ/mol energy window. Geometry reoptimizations, torsional scans [B3LYP/6-31G(d) level in vacuo] such as TDDFT calculations were performed with Gaussian 09[9] using various functionals (B3LYP, BH&HLYP, PBE0) and TZVP basis set. ECD spectra were generated as the sum of

Gaussians[10] with 3000 cm^{-1} half-height width (corresponding to ca. 31 nm at 320 nm), using dipole-velocity computed rotational strengths. Boltzmann distributions were estimated from the ZPVE corrected B3LYP/6-31G(d) energies. The MOLEKEL[11] software package was used for visualization of the results.

References

1. Debbab, A.; Aly, A. H.; Edrada-Ebel, R.; Wray, V.; Muller, W. E. G.; Totzke, F., Zirrgiebel, U.; Schälchtele, C.; Kubbutat, M. H. G.; Lin, W; Mosaddak, M.; Hakiki, A.; Proksch, P.; Ebel, R. *J. Nat. Prod.* **2009**, *72*, 626-631.
2. CLSI. Methods for Dilution Antimicrobial Susceptibility Tests for Bacteria That Grow Aerobically; Approved Standart – Ninth Edition. CLSI document M07-A9. Wayne, PA: Clinical and Laboratory Standarts Institute; 2012.
3. Burkholder, P. R.; Giles, N. H. *Am. J. Bot.* **1947**, *34*, 345–348.
4. Hiramatsu, K.; Hanaki, H.; Ino, T.; Yabuta, K.; Oguri, T.; Tenover, F. C. *J. Antimicrob. Chemother.* **1997**, *40*, 135-136.
5. Klare, I.; Heier, H.; Claus, H.; Boehme, G.; Marin, S.; Seltmann, G.; Hakenbeck, R.; Antanassova, V.; Witte, W. *Microb. Drug Resist.* **1995**, *1*, 265-272
6. Henrichfreise, B.; Wiegand, I.; Sherwood, K. J.; Wiedemann, B. *Antimicrob. Agents Chemother.* **2005**, *49*, 1668-1669.
7. Bara, R.; Zerfass, I.; Aly, A. H.; Goldbach-Gecke, H.; Raghavan, V.; Sass, P.; Mandi, A.; Wray, V.; Polavarapu, P. L.; Pretsch, A.; Lin, W.; Kurtan, T.; Debbab, A.; Broetz-Oesterhelt, H.; Proksch, P. *J. Med. Chem.* **2013**, *56*, 3257-3272.
8. MacroModel, Schrödinger LLC, 2012.
 http://www.schrodinger.com/productpage/14/11/
9. Frisch, M. J.; Trucks, G. W.; Schlegel, H. B.; Scuseria, G. E.; Robb, M. A.; Cheeseman, J. R.; Scalmani, G.; Barone, V; Mennucci, B.; Petersson, G. A.; Nakatsuji, H.; Caricato, M.; Li, X.; Hratchian, H. P.; Izmaylov, A. F.; Bloino, J.; Zheng, G.; Sonnenberg, J. L.; Hada, M.; Ehara, M.; Toyota, K.; Fukuda, R.; Hasegawa, J.; Ishida, M.; Nakajima, T.; Honda, Y.; Kitao, O.; Nakai, H.; Vreven, T.; Montgomery, J. A.; Peralta, J. E. Jr.; Ogliaro, F.; Bearpark, M.; Heyd, J. J.; Brothers, E.; Kudin, K. N.; Staroverov, V. N.; Kobayashi, R.; Normand, J.; Raghavachari, K.; Rendell, A.; Burant, J. C.; Iyengar, S. S.; Tomasi, J.; Cossi, M.; Rega, N.; Millam, J. M.; Klene, M.; Knox, J. E.; Cross, J. B.; Bakken, V.; Adamo, C.; Jaramillo, J.; Gomperts, R.; Stratmann, R. E.; Yazyev, O.; Austin, A. J.; Cammi, R.; Pomelli, C.; Ochterski, J. W.; Martin, R. L.; Morokuma, K.; Zakrzewski, V. G.; Voth, G. A.; Salvador, P.; Dannenberg, J. J.; Dapprich, S.; Daniels, A. D.; Farkas, O.; Foresman, J. B.; Ortiz, J. V.; Cioslowski, J.; Fox, D. J. Gaussian 09, Revision B.01, 2010, Gaussian, Inc., Wallingford CT.
10. Stephens, P. J.; Harada, N. *Chirality* **2010**, *22*, 229–233.

11. Varetto, U. MOLEKEL 5.4., 2009, Swiss National Supercomputing Centre: Manno, Switzerland.

Chapter 4

Dihydroanthracenone Metabolites from the Endophytic Fungus *Diaporthe melonis* Isolated from *Anonna squamosa*

Submitted to "Tetrahedron Letters"

Impact Factor: 2.397,

The overall contribution to the paper: 80% of the first author. The first author involved to all laboratory works as well as the manuscript preparation.

Dihydroanthracenone Metabolites from the Endophytic Fungus *Diaporthe melonis* Isolated from *Anonna squamosa*

Antonius R. B. Ola[a,b], Abdessamad Debbab[a], Tibor Kurtán[c], Heike Brötz-Oesterhelt[a], Amal H. Aly[a]*, Peter Proksch[a]*

[a] Institut für Pharmazeutische Biologie und Biotechnologie, Heinrich-Heine-Universität, Universitätsstrasse 1, 40225 Düsseldorf, Germany

[b] Department of Chemistry, Faculty of Science and Engineering, Nusa Cendana University, Jalan Adisucipto Penfui, 85001 Kupang, Indonesia

[c] Department of Organic Chemistry, University of Debrecen, POB 20, 4010 Debrecen, Hungary

---

* Corresponding authors. Tel.: +49-211-81-14173; e-mail: amal.hassan@uni-duesseldorf.de (A.H.A.). Tel.: +49-211-81-14163; fax: +49-211-81-11923; e-mail: proksch@uni-duesseldorf.de (P.P.).

Abstract

Chemical investigation of the endophytic fungus *Diaporthe melonis*, isolated from *Anonna squamosa*, yielded two new dihydroantharacenone atropodiastereomers, diaporthemins A (**1**) and B (**2**), together with the known flavomannin-6,6'-di-*O*-methyl ether (**3**). The structures of the new compounds were established on the basis of extensive 1D and 2D NMR spectroscopy, as well as by high resolution mass spectrometry and by CD spectroscopy. Compounds **1-3** were tested for their antimicrobial activity against a multi-resistant clinical isolate of *Staphylococcus aureus* 25697, a susceptible reference strain of *S. aureus* ATCC 29213 and against *Streptococcus pneumoniae* ATCC 49619. Compound **3** strongly inhibited *S. pneumonia* growth with a MIC value of 2 µg/mL and showed moderate activity against the *S. aureus* multi-resistant clinical isolate and susceptible reference strain (MIC 32 µg/mL), whereas **1** and **2** were not active against the tested strains.

Keywords: *Anonna squamosa*, Antibacterial activity, Dihydroanthracenones, *Diaporthe melonis*, Endophytic fungus.

Introduction

Endophytic fungi have proven to be important sources of new natural products with promising biological and pharmacological activities.[1-3] As part of our ongoing research focusing on the discovery of new natural products from fungal endophytes, we recently investigated *Diaporthe melonis* (syn. *Phomopsis cucurbitae*), an endophytic fungus isolated from *Anonna squamosa* (Annonaceae) collected in Kupang, eastern Indonesia. *A. squamosa* has been reported to produce various acetogenins with broad ranges of biological activity such as cytotoxic, antitumor, antiparasitic, pesticidal, antimicrobial, and immunosuppressive activities.[4-6] Traditionally, the plant has been used as an insecticidal agent. The seeds of *A. squamosa* had also been used in South China as a folk medicine to treat "malignant sore".[7]

The pleomorphic genus *Diaporthe* (anamorph *Phomopsis*) has been previously shown to produce bisanthraquinones, including (+)-epicytoskyrin, cytoskyrin A, (+)-rugolosin and (+)-1,1'-bislunatin.[8] In this study, we report the isolation and structure elucidation of two dihydroanthracenone atropodiastereomers, diaporthemins A (**1**) and B (**2**), as well as the known bisdihydroanthracenone flavomannin-6,6'-di-*O*-methyl ether FDM-A$_1$ (**3**) from *Diaporthe melonis* (Figure 1).

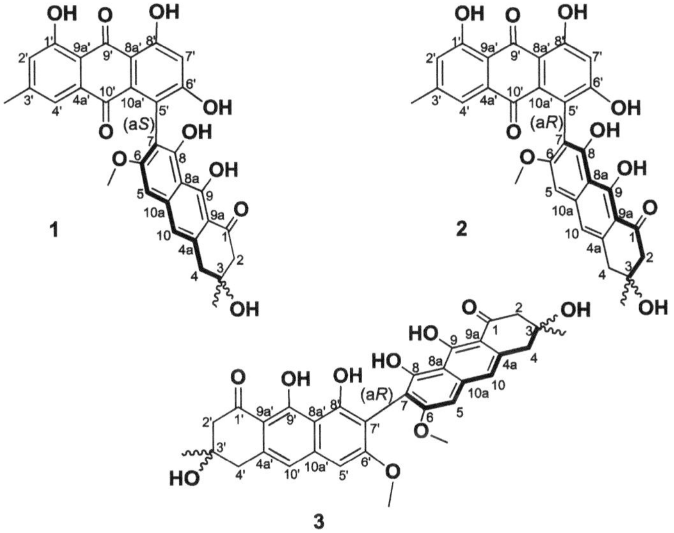

Figure 1. Structures of **1-3**.

Results and Discussion

The ethyl acetate extract from the culture of *D. melonis* on solid rice medium was subjected to chromatographic separation on different stationary phases, including silica gel and Sephadex LH-20 followed by final purification using semipreparative HPLC. Compound **1** was isolated as a brownish yellow amorphous powder. Its UV spectrum showed three absorption maxima λ_{max} (MeOH) at 223.7, 275.7 and 425 nm. This spectral pattern resembled those of talaromannin A and B,[9] hence suggesting that **1** bears close structural similarity to both compounds. Positive and negative ESIMS showed quasi-molecular ion peaks at *m/z* 557.2 [M+H]$^+$ and *m/z* 555.3 [M-H]$^-$, respectively, indicating a molecular weight of 556 g/mol with an increase of 14 mass units compared to talaromannin A and B.[9] The molecular formula of **1** was determined as $C_{31}H_{24}O_{10}$ on the basis of the prominent signal detected at *m/z* 557.14383 [M+H]$^+$ in the HRESIMS.

Analysis of the NMR data of **1** (Table 1) revealed the presence of the characteristic signals of a torosachrysone[10] and emodin building blocks, indicating that **1** is a heterodimer comprised of both units. The emodin moiety in **1** was verified through signals attributed to an aromatic methyl group at δ_H 2.34 ppm (3'-CH$_3$), as well as three aromatic methine protons at δ_H 6.89, 7.03 and 7.37 ppm (H-7', H-2' and H-4', respectively). Signals corresponding to the torosachrysone[10] unit in the ^{1}H NMR spectrum included a tertiary methyl group at δ_H 1.48 ppm (3-CH$_3$), two methylene groups at δ_H 2.87 and 3.11 ppm (CH$_2$-2 and CH$_2$-4, respectively), two aromatic proton singlets at δ_H 6.73 and 7.00 ppm (H-5 and H-10, respectively), and a methoxy group at δ_H 3.79 ppm (6-OCH$_3$), which accounted for the molecular weight increase in comparison to talaromannins A and B.

Table 1. NMR spectroscopic data of **1** and **2** at 500 (^{1}H) and 125 (^{13}C) MHz (δ in ppm, J in Hz)

Position	1				2[a]			
	δ_H[a]	ROESY[b]	HMBC[b]	δ_C[b,c]	δ_H	ROESY	HMBC	δ_C[c]
1				204.1				203.7
2	2.84, d (17.9)	3-CH$_3$	1, 3, 4, 3'-CH$_3$	51.5	2.85, d (17.9)	3-CH$_3$		51.3
	2.89, d (17.9)				2.90, d (17.9)			
3				70.6				71.0
4	3.11, br s	3-CH$_3$, 10	2, 3, 4a, 9a, 10	43.5	3.11, br s	3-CH$_3$, 10		43.8
4a				137.5				137.5
5	6.73, s	6-OCH$_3$, 10	6, 7, 8a, 10	99.0	6.73, s	6-OCH$_3$	8a, 10	98.9
6				162.2				161.9
7				108.7				108.3
8				155.8				156.8
8a				110.3				108.9
9				166.6				166.5
9a				108.3				110.5
10	7.00, s	4, 5	4, 5, 8a, 9a, 10a	118.0	7.00, s	4	4, 5, 8a	117.9
10a				141.6				141.5
1'				162.2				162.4
2'	7.03, s	3'-CH$_3$, 1'-OH	1', 4', 9a', 3'-CH$_3$	123.8	7.03, s	3'-CH$_3$	4', 9a', 3'-CH$_3$	123.5
3'				148.5				148.1
4'	7.37, s	3'-CH$_3$	2', 3'-CH$_3$, 9a', 10'	121.0	7.38, s	3'-CH$_3$	2', 3'-CH$_3$, 4a', 9a', 10'	121.2
4a'				137.6				137.5
5'				118.5				116.1
6'				162.5				162.5
7'	6.90, s	6'-OH, 8'-OH	5', 8', 8a'	108.3	6.89, s		5', 6, 8, 8a'	108.6
8'				165.5				165.0
8a'				110.7				111.6

Dihydroanthracenones from *D. melonis*

Position	δH (CDCl₃)[a]	HMBC	δC	δH (acetone-d₆)[b]	HMBC	δC
9'						113.2
9a'						182.9
10'			191.0			
10a'			113.7			
			182.6			
			136.0			
3-CH₃	1.48, s	2, 4	29.1	1.48, s	2, 3, 4	29.0
3'-CH₃	2.34, s	2', 4'	21.7	2.34, s	2', 3', 4'	21.7
6-OCH₃	3.79, s	5	55.8	3.78, s	6	56.1
8-OH	10.03, s	7, 8, 8a		10.05, s	7, 8, 8a	
9-OH	16.11, s			16.10, s		
1'-OH	12.08, s	2'	1', 2', 9a'	12.09, s	1', 2', 9a'	
6'-OH						
8'-OH	12.84, s	7'	7', 8', 8a'	12.84, s	7'	7', 8', 8a'

[a] In chloroform-d
[b] In aceton-d₆
[c] Derived from HSQC and HMBC experiments.

The structure of **1** was confirmed by thorough inspection of the COSY and ROESY spectra (Figure 2) showing correlations of 3'-CH$_3$ to the *meta*-coupled protons H-2' and H-4', and thus confirming the positon of 3'-CH$_3$ between both protons in the emodin unit. Similarly, 3-CH$_3$ correlated with CH$_2$-2 and CH$_2$-4, and the latter correlated with H-10, which in turn correlated with H-5. This together with the HMBC correlations (Figure 2) of H-10 to C-4, C-5, C-8a and C-9a, and of H-5 to C-7, C-8a and C-10, indicated a link of the torosachrysone moiety to the emodin part at C-7. The position of the methoxy group in the torosachrysone unit was established based on its HMBC correlation to C-6, as well as its ROESY correlation to the aromatic proton H-5. Further HMBC correlations were observed for H-7' to C-5', C-8' and C-8a', and for 8'-OH to C-7', C-8' and C8a'. Accordingly, the attachment of the emodin moiety to the torosachrysone part was established to be at position C-5'. The planar structure of **1** was further corroborated by comparison of its NMR data with those reported for talaromannins A and B,[9] skyrin,[10,11] emodin[12,13] and torosachrysone.[10]

Hence, **1** was identified as a new natural heterodimer and given the name diaporthemin A. It is worth mentioning that hitherto reported structural analogues bear either C5-C5',[14,15] C5-C7',[16] C7-C7',[9,17] C10-C5'[18] or C10-C7'[16,19-22] junctions between their monomeric subunits, thus **1** represents the first report of a C7-C5' linkage for this type of compounds.

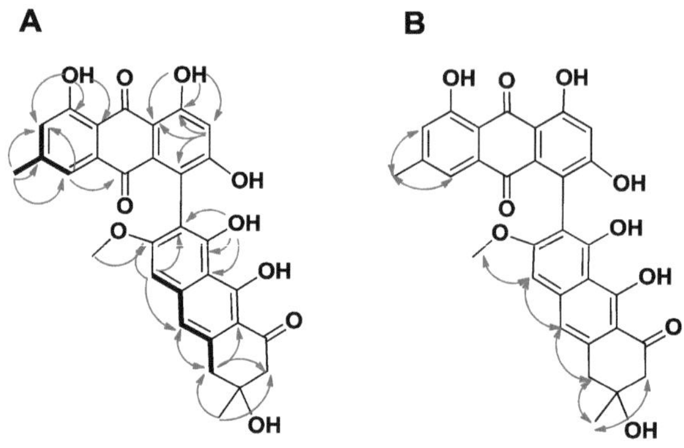

Figure 2. Important COSY (—), HMBC (→) (**A**) and ROESY (**B**) correlations of **1** and **2**.

The axial chirality of this type of molecules has been intensively discussed in the literature.[9,14-22] According to the exciton chirality method, a bisignate CD Cotton effect couplet centered around 275 nm, which results from the exciton coupling between two extended asymmetric aromatic chromophores, directly correlates with their helical twist. Consequently, an anticlockwise helical twist (negative chirality) has been suggested to exhibit a negative Cotton effect at longer wavelength and a positive one at shorter wavelength (A-type curve). In contrast, a CD spectrum with the mirror image Cotton effect couplet (B-type curve) indicates a clockwise helical twist and thus positive chirality.[14-18,20-22,25,26]

Hence, the axial chirality of **1** was tentatively determined by analysis of its CD spectrum, which showed a positive Cotton effect at shorter wavelength and a negative one at around 273 nm (A-type, Figure 3). These data suggested an anticlockwise helical twist between the aromatic chromophores and a positive $\omega_{C6',C5',C7,C8}$ dihedral angle, thus allowing the determination of the axial chirality as (aS)-**1**.

Dihydroanthracenones from *D. melonis*

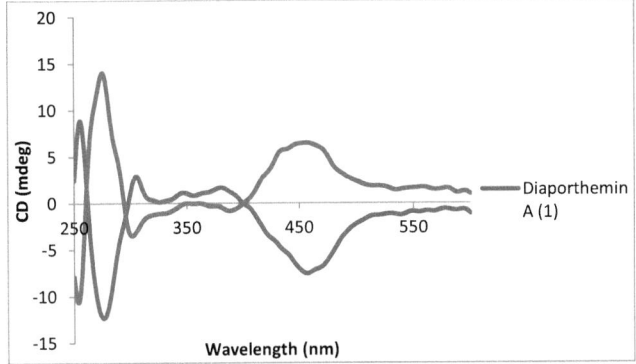

Figure 3. CD spectra of diaporthemins A (**1**) and B (**2**). The data were obtained in acetonitrile (0.7 mg/mL for both compounds) on the average of three scans corrected by subtracting a spectrum of the appropriate solution in the absence of the samples recorded under identical conditions. Each scan in the range 250-600 nm was obtained by taking points every 0.4 nm with an integration time of 0.4 s and a 1.0 nm bandwidth.

Compound **2** was isolated as a brown amorphous powder. It shared the same molecular weight (556 g/mol) and formula ($C_{31}H_{24}O_{10}$) with **1** as indicated by HRESIMS (*m/z* 557.14483 [M+H]$^+$). The UV spectrum of **2** showed almost identical absorption maxima (λ_{max} MeOH) to those observed for **1** (223.6, 275.7 and 423 nm). However, **2** was eluted ca. 1 min later than **1** upon HPLC analysis (R_t 31.0 and 31.9 min for **1** and **2**, respectively). Moreover, ^{1}H NMR and ^{13}C NMR data of both compounds were almost identical (Table 1). Analysis of 2D NMR spectra of **2** established the same planar structure as for **1** (Figure 2), thus suggesting that both compounds are stereoisomers differing either in the configuration of the biaryl axis or in the C-3 chirality center. As the CD features of these compounds are mainly determined by their axial chirality,[9,23,24] the mirror image CD spectra of **1** and **2** (Figure 3) suggested them to be atropodiastereomers differing in the axial chirality. Thus, **2** was identified as a new natural product and the name diaporthemin B was proposed. By analogy to 1, the negative Cotton effect at shorter wavelength and the positive one at ca. 273 nm (B-type, Figure 3) indicated a clockwise helical twist and a negative $\omega_{C6',C5',C7,C8}$ dihedral angle, and hence (aR) axial chirality for 2.

Compound 3 was identified on the basis of its NMR and mass spectrometric data, as well as by comparison with published data as flavomannin-6,6'-di-O-methyl ether FDM-A$_1$.[25,26] The CD spectrum of 3 exhibited an exciton couplet centered at ca. 280 nm (A-type, Figure 4) with positive and negative Cotton effects at shorter and longer wavelength, respectively, which indicated a positive $\omega_{C6',C7',C7,C8}$ dihedral angle and (aR) axial chirality for 3.

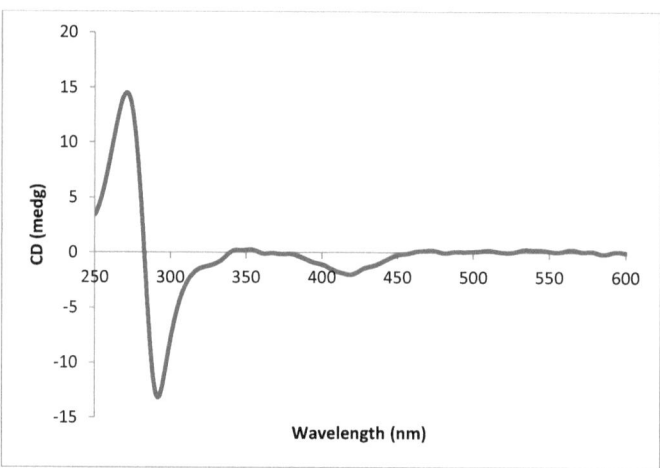

Figure 4. CD spectrum of flavomannin-6,6'-di-O-methyl ether FDM-A1 (3). The data were obtained in acetonitrile (1 mg/mL) on the average of three scans corrected by subtracting a spectrum of the appropriate solution in the absence of the samples recorded under identical conditions. Each scan in the range 250-600 nm was obtained by taking points every 0.5 nm with an integration time of 0.5 s and a 2 nm bandwidth.

The isolated compounds were tested for their antimicrobial activities against a multi-resistant clinical isolate of *Staphylococcus aureus* 25697, a susceptible strain of *S. aureus* ATCC 29213, and *Streptococcus pneumoniae* ATCC 49619. Only **3** showed strong growth inhibition against *S. pneumonia* with a MIC value of 2 µg/mL. Moderate activity against the multi-resistant clinical isolate *S. aureus* 25697 and the susceptible reference *S. aureus* (MIC 32 µg/mL) were also observed for **3**. In contrast, **1** and **2** did not show antibacterial activities against the tested strains up to a concentration of 64 µg/mL. Considering the fact that in a previous study the C7-C7' linked structurally related derivatives talaromannins A and B exhibited significant activities against multi-resistant *S. aureus*,[9] a C7-C5' linkage as in **1** and **2** seems to abolish antibacterial activity.

Acknowledgment

A.R.B.O. wishes to thank the DAAD for the doctoral fellowship. We would like also to thank Robin Visse (Organic Chemistry, HHU Duesseldorf) for the CD measurement, Dhana Tommy and Heike Goldbach-Gecke for the antibacterial assays. Support of the BMBF to P.P. is gratefully acknowledged.

Supplementary Data

Supplementary data (experimental section and compound characterization, as well as HPLC chromatograms, UV spectra, ^{1}H NMR spectra and HRESIMS of **1** and **2**) associated with this article can be found in the online version, at http://XXXXXXXXXXXXXX

References

1. Debbab, A.; Aly, A. H; Proksch, P. *Fungal Divers*. **2012**, *57*, 45-83.
2. Aly, A. H.; Debbab, A.; Proksch, P. *Appl. Microbiol. Biotechnol*. **2011**, *90*, 1829-1845.
3. Aly, A. H; Debbab, A.; Kjer, J.; Proksch, P. *Fungal Divers*. **2010**, *41*, 1-16.

4. Alali, F. Q.; Liu, X.-X.; McLaughlin, J. L. *J. Nat. Prod.* **1999**, *62*, 504-540.

5. Bermejo, A.; Figadere, B.; Zafra-Polo, M.-C.; Barrachina, I.; Estornell, E.; Cortes, D. *Nat. Prod. Rep.* **2005**, *22*, 269-303.

6. Araya, H.; Sahai, M.; Singh, S.; Singh, A. K.; Yoshida, M.; Hara, N.; Fujimoto, Y. *Phytochemistry.* **2002**, *61*, 999-1004.

7. Chen, Y.; Chen, J.-w.; Wang, Y.; Xu, S.-s.; Li, X. *Food Chem.* **2012**, *135*, 960-966.

8. Agusta, A.; Ohashi, K.; Shibuya, H. *Chem. Pharm.Bull.* **2006**, *54*, 579-582.

9. Bara, R.; Zerfass, I.; Aly, A. H.; Goldbach-Gecke, H.; Raghavan, V.; Sass, P.; Mandi, A.; Wray, V.; Polavarapu, P. L.; Pretsch, A.; Lin, W.; Kurtan, T.; Debbab, A.; Broetz-Oesterhelt, H.; Proksch, P. *J. Med. Chem.* **2013**, *56*, 3257-3272.

10. Gill, M.; Saubern, S.; Yu, J. *Aust. J. Chem.* **2000**, *53*, 213-220.

11. Vargas, F.; Rivas, C.; Zoltan, T.; Lopez, V.; Ortega, J.; Izzo, C.; Pineda, M.; Medina, J.; Medina, E.; Rosales, L. *Av. Quim (Avances en Quimica).* **2008**, *3*, 7-14.

12. Choi, J. S.; Jung, J. H.; Lee, H. J.; Kang, S. S. *Arch. Pharm. Res.* **1996**, *19*, 302-306.

13. Danielsen, K.; Aksnes, D. W. *Magn. Reson. Chem.* **1992**, *30*, 359-360.

14. Antonowitz, A.; Gill, M.; Morgan, P. M.; Yu, J. *Phytochemistry* **1994**, *37*, 1679-1683.

15. Gill, M.; Morgan, P. M. *ARKIVOC* **2004**, *10*, 152-165.

16. Kitanaka, S.; Takido, M. *Phytochemistry* **1982**, *21*, 2103-2106.

17. Gill, M.; Raudies, E.; Yu, J. *Aust. J. Chem.* **1999**, *52*, 989-992.

18. Buchanan, M. S.; Gill, M.; Millar, P.; Phonh-Axa, S.; Raudies, E.; Yu, J. *J. Chem. Soc., Perkin Trans. 1* **1999**, *7*, 795-802.

19. Alemayehu, G.; Abegaz, B.; Kraus, W. *Phytochemistry* **1998**, *48*, 699-702.

20. Elsworth, C.; Gill, M.; Gimenez, A.; Milanovic, N. M.; Raudies, E. *J. Chem. Soc., Perkin Trans. 1* **1999**, *2*, 119-125.

21. Mueller, M.; Lamottke, K.; Steglich, W.; Busemann, S.; Reichert, M.; Bringmann, G.; Spiteller, P. *Eur. J. Org. Chem.* **2004**, 4850-4855

22. Steglich, W.; Toepfer-Petersen, E.; Pils, I. *Z. Naturforsch C* **1973**, *28*, 354-355.

23. Debbab, A.; Aly, A. H.; Edrada-Ebel, R.; Wray, V.; Pretsch, A.; Pescitelli, G.; Kurtan, T.; Proksch, P. *Eur. J. Org. Chem.* **2012**, 1351-1359.

24. Polavarapu, P. L.; Jeirath, N.; Kurtán, T.; Pescitelli, G.; Krohn, K. *Chirality* **2009**, *21*, E202-E207.

25. Gill, M.; Morgan, P. M. *ARKIVOC* **2001**, *7*, 145-156.

26. Gill, M.; Giménez, A.; Jhingran, A. G.; Palfreyman, A. R. *Tetrahedron: Asymmetry* **1990**, *1*, 621-634.

Supporting Information

Dihydroanthracenone Metabolites from the Endophytic Fungus *Diaporthe melonis* Isolated from *Anonna squamosa*

Antonius Ola[a,b], Abdessamad Debbab[a], Tibor Kurtán[c], Heike Brötz-Oesterhelt[a], Amal H. Aly[a]*, Peter Proksch[a]*

[a] Institut für Pharmazeutische Biologie und Biotechnologie, Heinrich-Heine-Universität, Universitätsstrasse 1, 40225 Düsseldorf, Germany

[b] Department of Chemistry, Faculty of Science and Engineering, Nusa Cendana University, Jalan Adisucipto Penfui, 85001 Kupang, Indonesia

[c] Department of Organic Chemistry, University of Debrecen, POB 20, 4010 Debrecen, Hungary

Table of Contents

Experimental section	S3
References	S6
HPLC chromatogram and UV spectrum of **1**	S7
HRESIMS of **1**	S8
^{1}H NMR spectrum of **1**	S9
HPLC chromatogram and UV spectrum of **2**	S10
HRESIMS of **2**	S11
^{1}H NMR spectrum of **2**	S12

Experimental Section

General Experimental Procedures. Optical rotations were determined on a Perkin-Elmer-241 MC polarimeter. 1D and 2D NMR spectra were recorded on an Avance DMX 600 NMR spectrometer. CD spectra were measured on a Jasco J-600 instrument. Mass spectra (ESIMS) were measured on a LCMS HP1100 Agilent Finnigan LCQ Deca XP Thermoquest and high-resolution electrospray ionization mass spectroscopy (HRESIMS) were recorded on an UHR-TOF maXis 4G (Bruker Daltonics, Bremen) mass spectrometer. HPLC analysis was performed with a Dionex P580 system coupled to a photodiode array detector (UVD340S); routine detection was at 235, 254, 280, and 340 nm. The separation column (125 × 4 mm) was prefilled with Eurosphere-10 C_{18} (Knauer, Germany), and the following gradient was used (MeOH, 0.1% HCOOH in H_2O): 0-5 min (10% MeOH); 5-35 min (10-100% MeOH); 35-45 min (100% MeOH). Semi-preparative HPLC was performed using a Merck Hitachi HPLC System (UV detector L-7400; Pump L-7100; Eurosphere-100 C_{18}, 300 × 8 mm, Knauer, Germany). Column chromatography was performed on Silica gel 60 M (230−3400 mesh ASTM, Macherey-Nagel GmbH & Co. KG, Dueren, Germany) and Sephadex LH-20 (Sigma). TLC was carried out on precoated silica gel plates (silica gel 60 F-254, Merck KGaA, Darmstadt, Germany) for monitoring of fractions using EtOAc/MeOH/H_2O (30:5:4) and CH_2Cl_2/MeOH (9:1) as solvent systems. Detection was at 254 and 366 nm or by spraying the plates with anisaldehyde reagent. Solvents were distilled before use, and spectral grade solvents were used for spectroscopic measurements.

Plant and Fungal Material. The endophytic fungus was isolated from inner tissues of fresh healthy leaves of *A. squamosa*, collected in February 2012 at Kupang, eastern Indonesia, according to previously described methods.[1] A voucher strain (strain designation AB 1.1) is kept in the authors' laboratory (P.P.).

Identification of the Fungal Strain. The fungal strain was identified according to a molecular biological protocol by DNA amplification and sequencing of the ITS region as described previously.[1,2] The BLAST search result showed that the sequence had 98% similarity to the sequence of *Diaporthe melonis*. Sequence data have been submitted to GenBank with accession number KF493656.

Cultivation. Three Erlenmeyer flasks (1L each) containing 100 g of commercially available rice and 100 mL of distilled water were autoclaved. A small part of the medium from a Petri dish containing the pure fungal culture was transferred under sterile conditions to the solid medium and grown at room temperature (22 °C) for 21 days.

Extraction and Isolation. Cultures were exhaustively extracted with EtOAc and the dried crude extract (2.3 g) was subjected to vacuum liquid chromatography (VLC) using *n*-hexane/EtOAc followed by CH_2Cl_2/MeOH step gradient elution. Fraction 7, eluted with CH_2Cl_2/MeOH (9:1), was further purified using Sephadex LH-20 (CH_2Cl_2: MeOH, 1:1) followed by semi-preparative HPLC (CH_3CN/H_2O) to yield **1** (2.1 mg) and **2** (1.5 mg). Fraction 8, eluted with CH_2Cl_2/MeOH (7:3), was further purified by semi-preparative HPLC (CH_3CN/H_2O) to afford **3** (5 mg).

Diaporthemin A (1): brownish yellow amorphous powder; $[\alpha]_D^{21}$ -769.9 (*c*. 0.05, acetone); UV λ_{max} (PDA) 223.7, 275.7 and 425 nm; CD (CH_3CN, $c = 1.3 \times 10^{-3}$) λ_{max} ($\Delta\varepsilon$): 275sh (-3.0), 305 (0.77), 382 (0.4), 457sh (-1.86); ^1H and ^{13}C NMR (500 MHz, $CDCl_3$) see Table 1; ESIMS *m/z* 557.2 $[M+H]^+$, 555.3 $[M-H]^-$; HRESIMS *m/z* 557.14383 $[M+H]^+$ (calcd for $C_{31}H_{25}O_{10}$, 557.14422).

Diaporthemin B (2): brownish yellow amorphous powder; $[\alpha]_D^{21}$ +152.4 (c. 0.2, acetone); UV λ_{max} (PDA) 223.6, 275.7 and 423 nm; CD (CH$_3$CN, c = 1.3×10^{-3}) λ_{max} ($\Delta\varepsilon$): 275sh (3.4), 305 (−0.87), 382 (−0.3), 457sh (1.60); ^{1}H and ^{13}C NMR (500 MHz, CDCl$_3$) see Table 1; ESIMS m/z 557.2 [M+H]$^+$, 555.3 [M-H]$^-$; HRESIMS m/z 557.14483 [M+H]$^+$ (calcd for C$_{31}$H$_{25}$O$_{10}$, 557.14422).

Antibacterial Assay. MIC values were determined by the broth microdilution method according to CLSI guidelines.[3] For preparation of the inoculum, the direct colony suspension method was used with an inoculum of 5 x 10^5 colony forming units/mL after the last dilution step. The bacterial strain panel included antibiotic susceptible CLSI (Clinical Laboratory Standard Institute) quality control strains: *Staphylococcus aureus* ATCC 29213 and *Streptococcus pneumoniae* ATCC 49619, as well as the (multi)drug-resistant clinical isolate *Staphylococcus aureus* 25697 (AiCuris, Wuppertal, Germany).

References

1. Kjer, J.; Wray, V.; Edrada-Ebel, R.; Ebel, R.; Pretsch, A.; Lin, W.; Proksch, P. *J. Nat. Prod.* **2009**, *72*, 2053-2057.

2. Debbab, A.; Aly, A. H.; Edrada-Ebel, R.; Wray, V.; Mueller, W. E. G.; Totzke, F.; Zirrgiebel, U.; Schaechtele, C.; Kubbutat, M. H. G.; Lin, W. H.; Mosaddak, M.; Hakiki, A.; Proksch, P.; Ebel, R. *J. Nat. Prod.* **2009**, *72*, 626-631.

3. CLSI. Methods for Dilution Antimicrobial Susceptibility Tests for Bacteria That Grow Aerobically; Approved Standard – Ninth Edition. CLSI document M07-A9. Wayne, PA: Clinical and Laboratory Standards Institute; 2012.

HPLC chromatogram and UV spectrum of **1**

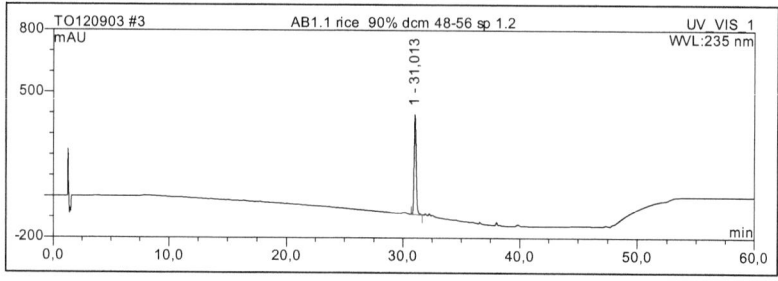
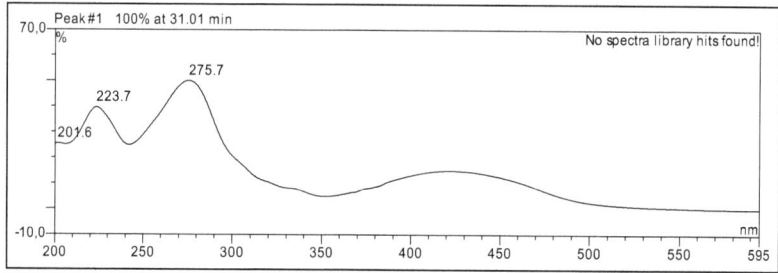

HRESIMS of **1**

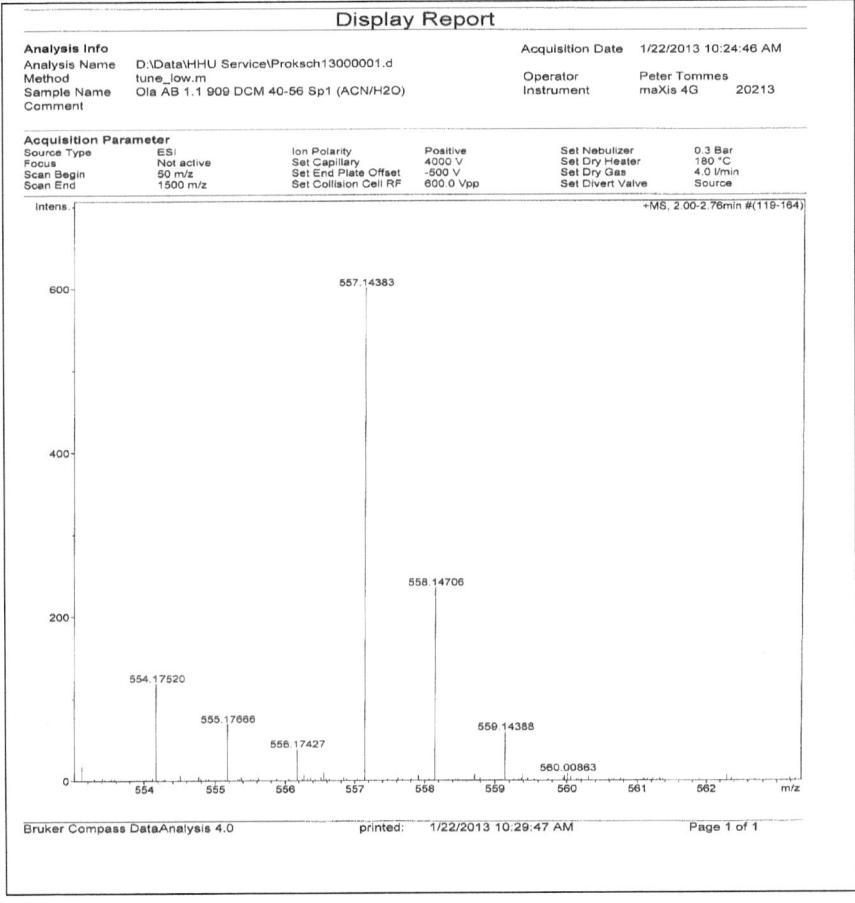

^{1}H NMR spectrum of **1** in chloroform-*d* (500 MHz)

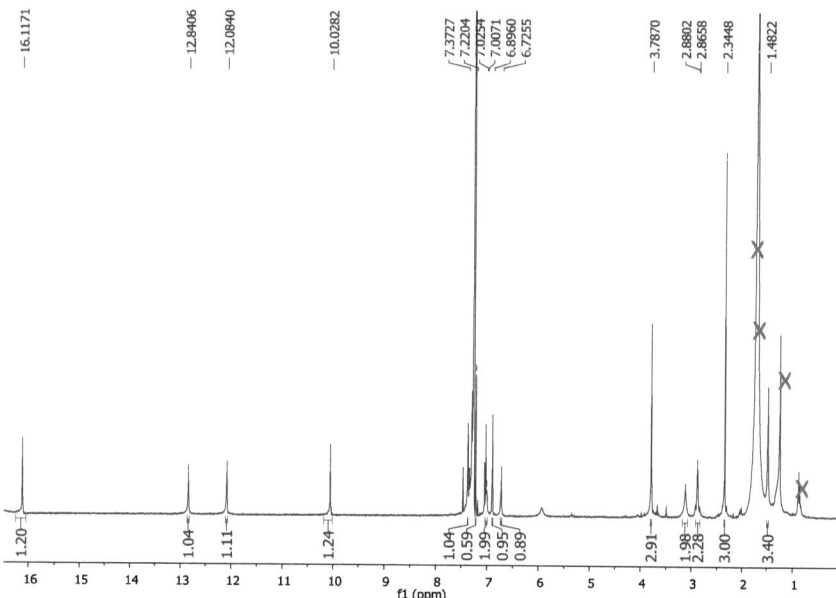

HPLC chromatogram and UV spectrum of **2**

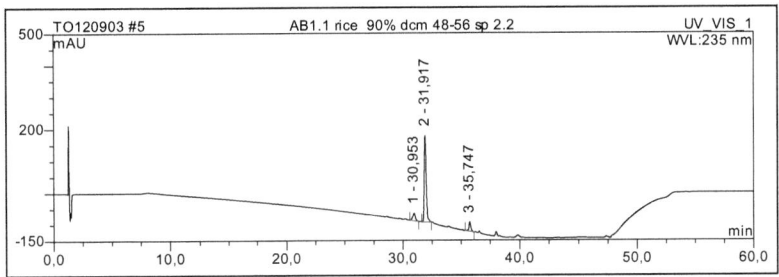

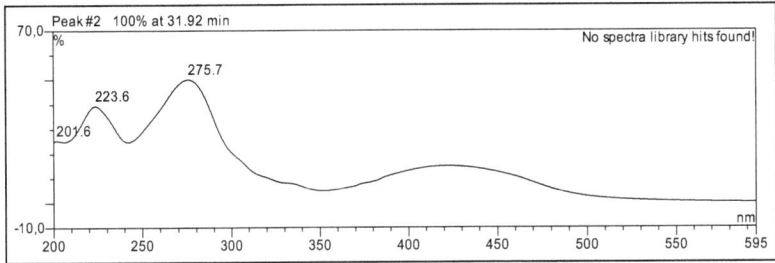

HRESIMS of **2**

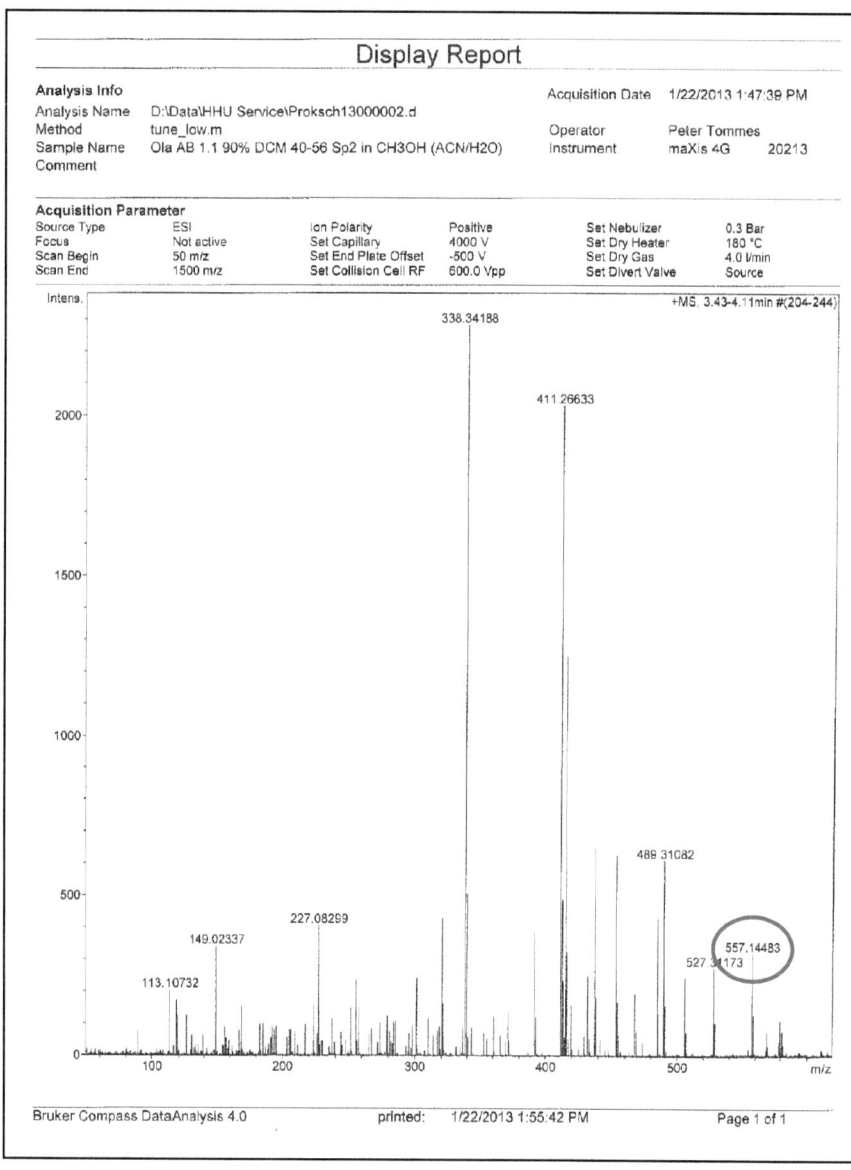

¹H NMR spectrum of **2** in chloroform-*d* (500 MHz)

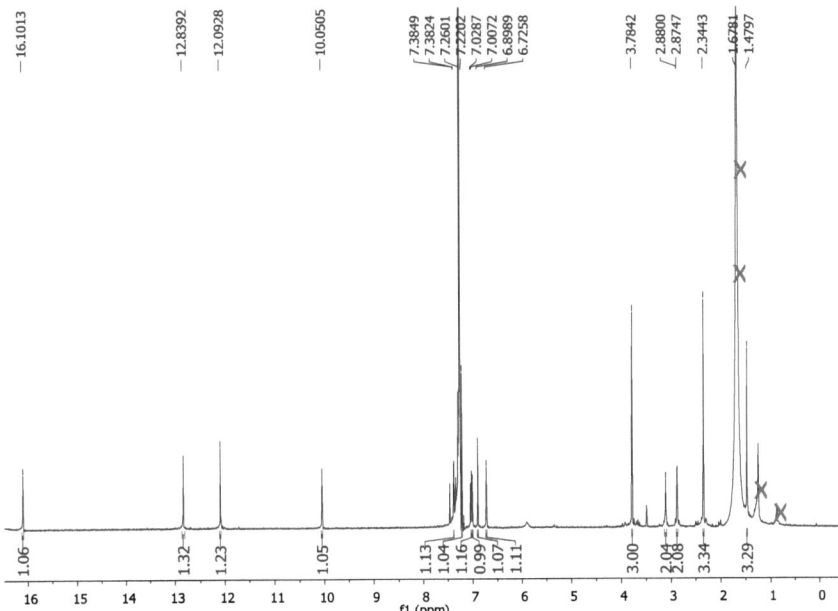

Chapter 5

Discussion

5.1 New Approaches for Activation of Silent Biosynthetic Pathways in Fungi

5.1.1 Induction of Silent Biosynthetic Pathways in the Endophytic Fungus *Fusarium tricinctum* through Coculture-An Ecological Perspective.

Fungi are known to synthesize a multitude of low-molecular-mass compounds known as secondary metabolites. Many of these compounds have important applications such as antibiotics, cholesterol-lowering agents, tumor inhibitors and immunosuppressants for organ transplantations. Recent advanced studies have shown that the potential of fungi to produce bioactive molecules has not been substantially maximized due to many fungal secondary metabolite biosynthesis pathways remaining silent under standard cultivation conditions. Moreover, the usual ways of screening for secondary metabolites that are produced by microorganisms in response to variation in the medium, pH, temperature, aeration, light, and so on are not sufficient if the physiological and/or ecological triggers for activating silent biosynthetic pathways are not known (Brakhage, 2013). As a result, only a minority of potential chemical compounds is produced when culturing microbes under standardized laboratory conditions (Scherlach and Hertweck, 2009; Brakhage and Schroeckh, 2011). Therefore, the activation of silent biosynthetic pathway is crucially needed to maximize the chemical diversity of endophytic fungi.

Although several approaches have been applied such as molecular-based techniques (genetic engineering) and manipulation of chromatin modification (the use of epigenetic modifiers), strategies based on an ecological perspective, namely simulation of natural habitats is of special interest. Simulation of natural habitat such as coculture of two or more different microorganism is based on the assumption of that every natural product is produced as a result of the interaction of an organism with its natural environment. Particularly, microorganism might use these natural products for intra- and interspecies cross talk. The idea of an interspecies cross talk leading to chemical diversity has also been applied in our laboratory through cocultures between fungus and bacteria (*F. tricinctum* vs *Bacillus subtilis*, *F. tricinctum* vs *Streptomyces lividans* and *F. tricinctum* vs *S. coelicolor*) (publication 1) and also fungus (*F. tricinctum* vs *Chaetomium aureus* and *F. tricinctum* vs *Talaromyces*

Discussion

wortmanii). Interestingly, different strains induced the accumulation of different metabolites. More specifically, coculturing the endophytic fungus *F. tricinctum* with the bacterium *Bacillus subtilis* has led to production of new chemical metabolites and also to an increased accumulation of fungal antibiotic compounds (publication 1). As the bacterium *Bacillus subtilis* has been also reported as an endophyte from plants, the scenario of coculture here probably also occurs in the nature.

Therefore, simulation of environmental conditions such as coculture is an excellent approach to pave natural product drug discovery program. This is due to the fact that interaction of the microorganism in the natural habitat could activate the biosynthetic pathway to produce new metabolites as info chemicals or to respond to the complex interactions in the environment by the production of antibiotic metabolites to suppress the growth of inhibitors (publication 1). There are two possible ways for the activation of silent biosynthetic pathways in these mixed cocultures (Scherlach and Hertweck, 2009):

1. One organism stimulates assembly of natural products in the other organism *via* secretion of chemical signals or physical interaction

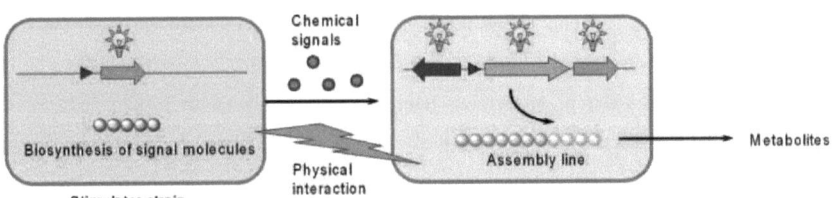

2. Exchange of chemical signals: One organism induces biosynthesis of signaling molecules which then stimulate production of cryptic metabolites.

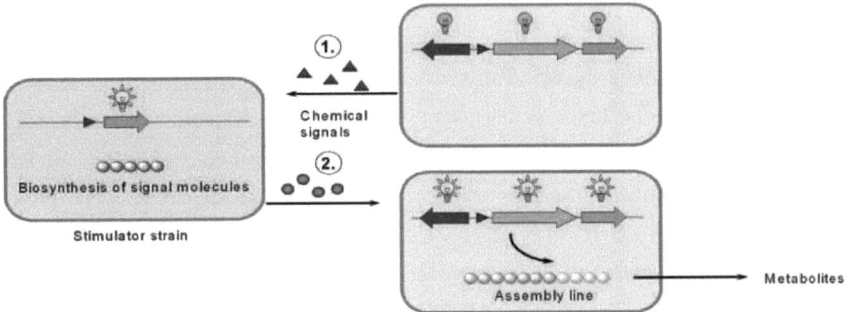

However, it has been also noticed that activation of silent biosynthetic pathways in fungi was not influenced by diffusible chemical signals but depends mainly on the intimate physical interaction between bacteria and fungi. Continuation of this work showed that the bacterium was able to trigger alterations of histone modification at the molecular level in the fungus. It had been demonstrated that the Saga/Ada complex containing the HAT GcnE and the AdaB protein is required for the induction of the orselinic gene cluster of *Aspergillus nidulans* by the bacterium *S. rapamycinicus* (Nützman *et al.*, 2011). It was suggested that specificity for activation of cluster genes depends on H3K9 acetylation. Furthermore, the addition of suberoylanilide hydroxamic acid (SAHA-HDAC inhibitor), activated orsellinic acid biosynthesis by *A. nidulans* without the need for co-incubation with *S. rapamycinicus*.

5.1.2 Induction of Silent Biosynthetic Pathways in the Endophytic Fungus *Fusarium tricinctum* with Epigenetic Modifiers.

Chromatin is the complex of DNA and the associated histone proteins. Histone proteins are made up of long chains of amino acids. If the amino acids are changed, the shape of the histone sphere might be modified and the chromatin structure might change as well. The unstructured N-termini of histones (called histone tails) are important features of the core histones and have been the target of various covalent modifications such as acetylation, methylation, phosphorylation, sumoylation and ubiquitylation.

Various histone modifications have been associated with the regulation of secondary metabolism gene clusters especially the acetylation of histones H3 and H4 and targeted lysine and arginine residues. The process of acetylating and deacetylating lysine is carried out by two sets of enzymes: histone acetyltransferases (HATs) and histone deacetylases (HDACs), respectively (Cichewicz, 2010). It has been demonstrated that disruption of histone deacetylase activity in *Aspergillus nidulans* led to the transcriptional activation of gene clusters responsible for the production of sterigmatocystin and penicillin (Shwab *et al.*, 2007). Based on this assumption and observation, it is rational to apply small molecules of chromatin-modulators such as inhibitors of histone acetyltransferases (HAT), histone deacetylases (HDACs) or DNA methyltransferases (DMATs) to access silent natural product pathways and to enhance the native production of fungal secondary metabolites.

Discussion

We applied the HDAC inhibitors suberoylanilide hydroxamic acid (SAHA) and valproic acid together with 5-azacythidine (DMAT inhibitor) to the fungus *F. tricinctum*. It is interesting to note that addition of HDAC inhibitors SAHA and valproic acid to liquid culture of *F. tricinctum* resulted in the enhanced production of the native secondary metabolites of the fungus up to four fold and also the induction of silent polyketide biosynthesis pathway to produce lateropyrone although in a low amount. Furthermore, treating the fungus with the DMAT 5-azacythidine only resulted in the enhanced production of native secondary metabolites of *F. tricinctum* including mainly the peptides enniatins and the lipopeptide fusaristatin A up to three fold. However, multiple treatments of HDAC inhibitors of SAHA and valproic acid and DMAT 5-azacythidine gave the opposite effect, i.e lower production of native secondary metabolites compared to the controls of the fungus.

These findings suggest that HDAC and DMAT inhibitors could be applied to maximize secondary metabolite production. Moreover, this technique does not require strain-dependent genetic manipulation and can thus be applied to any fungal strain. However, concentration of the inhibitors becomes an important issue of this technique. In contrast to the reports of the use of HDAC and DMAT inhibitors, HAT inhibitors to induce secondary metabolite production are rarely been used. Latest findings have shown that the HAT inhibitor anacardic acids blocks the activation of gene clusters in the fungi *A. nidulans* and *A. fumigatus* (Nützman *et al.*, 2011 and König *et al.*, 2013).

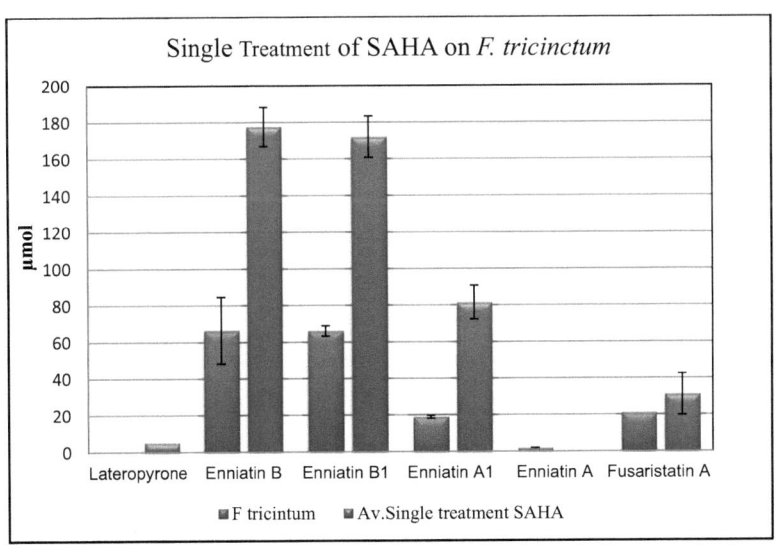

5.2 Atropisomer Natural Products from the Endophytic Fungi *Aspergillus fumigatiaffinis* and *Diaporthe melonis*

Atropisomers are stereoisomers resulting from hindered axial rotation that can, in principle, interconvert thermally, but for which the half-life of interconversion is ~1000 s (16.7 min) or longer, thus allowing analytical separation. Thus, the minimum free energy barrier $\Delta G^{\ddagger}$ required varies with temperature (e.g. $\Delta G^{\ddagger}_{200K}$ = 61.6 kJmol^{-1}, ΔG_{300K} = 93.5 kJmol^{-1}, and $\Delta G^{\ddagger}_{350K}$ = 109 kJmol^{-1}) (Claydon *et al.*, 2005). Atropisomers can significantly differ in biological activity, pharmacokinetics and toxicity. One of the best examples is (*S*)-gossypol as an antifertility agent and (*R*)-gossypol as an anticancer agent. Thus, atropisomers can have a significant impact on drug discovery and development process.

Several atropisomeric natural products such as the glycopeptide antibiotic vancomycin, skyrin, the bisanthracene derivatives flavomannins A-D, talaromannin A and B (Bara *et al.*, 2013), and alterporriol D (Debbab *et al.*, 2012) were reported to be active against a panel of pathogenic microorganisms including MRSA. During this study, the atropisomer neosartorin from *A. fumigatiaffinis* and diaporthemin A and B from *Diaporthe melonis* was isolated (publications 2 and 3).

5.2.1 The Influence of the Linkage and of the Substituents for the Biological Activity of Naturally Occurring Tetrahydroxanthone Atropisomers

Neosartorin is closely structurall related to the phomoxanthones A and B and to dicerandrols A-C. They are all structurally related to the secalonic acids. Neosartorin and phomoxanthone B are two examples of naturally occurring C2-C4' linkage of tetrahydroxanthone dimers. It is interesting to note that the linkage of the tetrahydroxanthone units is important for antibiotic activity and the cytotoxic activity. Phomoxanthone A having a C4-C4' linkage showed stronger antibiotic but also cytotoxic activity compared to phomoxanthone B with the connection at C2-C4' (Isaka et al., 2001). It is interesting to note that neosartorin having the biaryl axis at C2-C4' was inactive in the cytotoxicity assay, but showed strong antibiotic properties against several standard strains of Gram positive bacteria including the multiresistant clinical isolates (publication 2). Therefore, in term of directing the development of antibiotic activity, atropisomers with a C2-C4' are favoured while for the development of anticancer agents emphasis should be on the C4-C4' linkage as has been confirmed in a

Discussion

previous study (Rönsberg et al., 2013). Acetate group had been previously found to be important for the biological activity of phomoxanthone A, as the acetylation of phomoxanthone A abolished its biological activity (Isaka et al., 2001 and Rönsberg et al., 2013). Therefore, addition of acetate group to neosartorin probably contributes to the enhancement of antibiotic activity but probably also to its cytotoxicity.

Phomoxanthone A   Neosartorin

5.2.2 The Influence of the Linkage and of the Substituents for the Biological Activity of Naturally Occurring Dihydroanthracenone Atropisomers

Diaporthemin A and B are structurally related to the dihydroanthracenones talaromannin A and B which has been previously isolated from the endophytic fungi *Talaromyces wortmanii* (Bara et al., 2013). In the previous study, talaromannin A and B have shown antibiotic activity against several multiresistant clinical isolates. In addition, flavomannin A had also been reported to inhibit the growth of several multiresistant clinical isolates. Flavomannin A is structurally related to the likewise isolated flavomannin-6,6'-di-*O*-methyl ether FDM-A$_1$ in this study (publication 3). Flavomannin A and FDM-A$_1$ have the linkage of C7-C7' and showed antibacterial activity against several Gram positive bacteria (publication 3). However, flavomannin A and B were more active against MRSA than flavomannin-6,6'-di-*O*-methyl ether FDM-A$_1$, whereas FDM-A$_1$ was found to be more active against *Streptococcus pneumonia* compared to flavomannin A-D. This suggests that the addition of methoxy group might have no role to the antibacterial activity.

Furthermore, diaporthemin A and B were found to be antibiotically inactive up to a tested concentration of 64 μg/mL compared to the antibiotically active talaromannin A and B (Publication 3). As the addition of methoxy group had no influence on the antibacterial

Discussion

activity as had been observed for the first case with flavomannin-6,6'-di-*O*-methyl ether FDM-A$_1$, the different linkage of the atropisomers abolishs the biological activity as observed for diaporthemin A and B (publication 3). Therefore, the linkage of the monomers underlying the respective atropisomers plays a crucial role for the antibiotic activity.

Diaporthemin A   Talaromannin B

5.3 Biosynthetic Relationship of Polyketide Metabolites Isolated in This Study

The biosynthetic relationship between secondary metabolites of polyketide origin, including monomeric and dimeric tetrahyroxanthone and dihydroanthracenones derivative had been proposed via the acetate/malonate pathway (Gill *et al.*, 1989 and Kurobane *et al.*, 1979). The polyketides isolated in this study from *Diaporthe melonis* and *Aspergillus fumigatiaffinis* are probably derived from an octaketide which undergoes condensation and cyclisation to yield carboxylic acid **1** which can be either decarboxylated to pre anthraquinone precursors athrocrysone or dehydrated to the anthrone endocrine. Athrochrysone can either undergo water elimination and oxidation to yield the anthraquinone emodin or be methylated yielding torasachrysone (Gill *et al.*, 1989).

Torasachrysone and emodin would probably further undergo oxidative coupling followed by dimerization to yield diaporthemin A and B. Furthermore, emodin may be enzymatically transformed into benzophenones (**2** and **3**) by oxidative ring opening between C-4a and C-10. Benzophenones might undergo oxidative coupling and dimerization to yield neosartorin as proposed for the structurally related secalonic acids suggested by Kurobane *et al* (1979). The plausible biosynthetic pathway of the tetrahydroxanthone dimer neosartorin and dihydroanthracenones dimers diaporthemin A and B could be depicted in the following scheme.

Discussion

Scheme depicting the proposed biosynthetic pathway of diaporthemins and neosartorin

Athrocrysone

Torosachrysone

Emodin

Diaporthemin

Neosartorin

RERERENCES

Abraham, E. P., Chain, E., Fletcher, C. M., Gardner, A. D., Heatley, N. G., Jennings, M. A., Florey, H. W. (1941). Further observations on penicillin. *Lancet* **II**, 177-188.

Alberts, A. W. (1988). Discovery, biochemistry and biology of lovastatin. *Am. J. Cardiol.* **62**(15), 10j-15j.

Alberts, A. W. (1990). Lovastatin and simvastatin--inhibitors of HMG CoA reductase and cholesterol biosynthesis. *Cardiology*, 77 (4), 14-21.

Aly, A. H., Debbab, A., Proksch, P. (2011). Fifty years of drug discovery from fungi. *Fungal Divers.* **50**(1),3-19.

Aly, A. H.,Debbab, A., Proksch, P. (2011). Fungal endophytes: unique plant inhabitants with great promises. *Appl. Microbiol. Biotechnol.* **90**(6),1829-1845.

Aly, A. H., Debbab, A., Proksch, P. (2013). Fungal endophytes - secret producers of bioactive plant metabolites. *Pharmazie* **68**(7): 499-505.

Anke, T. (1995). The antifungal strobilurins and their possible ecological role. *Can. J. Bot.* **73(1),** S940-S945.

Arnold, A. E., Henk, D. A., Eells, R. L., Lutzoni, F., Vilgalys, R. (2007). Diversity and phylogenetic affinities of foliar fungal endophytes in loblolly pine inferred by culturing and environmental PCR. *Mycologia* **99**(2), 185-206.

Asai, T., Chung, Y-M., Sakurai, H., Ozeki, T.,Chang, F-R., Yamashita, K., Oshima, Y. (2011). Tenuipyrone, a Novel Skeletal Polyketide from the Entomopathogenic Fungus, *Isaria tenuipes*, Cultivated in the Presence of Epigenetic Modifiers. *Org.Lett.* 14(2), 513-515.

Bacon, C. W., White, J. F., Dekker, M. Jr. (2000). Microbial Endophytes. *Biodiver. Conserv.* **11**(4), 747-748.

Bara, R., Zerfass, I., Aly, A. H., Goldbach-Gecke, H., Raghavan, V., Sass, P., Mandi, A., Wray, V., Polavarapu, P. L., Pretsch, A., Lin, W., Kurtan, T., Debbab, A., Broetz-Oesterhelt, H., Proksch, P. (2013). Atropisomeric Dihydroanthracenones as Inhibitors of Multiresistant *Staphylococcus aureus*. *J. Med. Chem.* 56, 3257-3272.

Bergmann, S., Schuemann, J., Scherlach, K., Lange, C., Brakhage, A. A., Hertweck, C. (2007). Genomics-driven discovery of PKS-NRPS hybrid metabolites from *Aspergillus nidulans*. *Nat. Chem. Biol.* **3**(4), 213-217.

Bok, J. W. and Keller, N. P. (2004). LaeA, a regulator of secondary metabolism in *Aspergillus* spp. *Eukaryot Cell* **3**(2), 527-535.

Borchardt, J. K. (2002). The Beginnings of Drug Therapy: Ancient Mesopotamian Medicine. *Drug News Perspect.* **15**(3), 187-192.

Borchardt J. K. (2003). Beginnings of drug therapy: drug therapy in ancient India. *Drug News Perspect.* **16**(6), 403-408.

References

Brakhage Axel, A. (2013). Regulation of fungal secondary metabolism. *Nat. Rev. Microbiol.* **11**(1), 21-32.

Brakhage, A. A., Schroeckh, V. (2011). Fungal secondary metabolites – Strategies to activate silent gene clusters. *Fungal Genet. Biol.* **48**(1), 15-22.

Buckland, B.,Gbewonyo, K.,Hallada, T., Kaplan, L., Masurekar, P. (1989). Production of lovastatin, an inhibitor of cholesterol accumulation in humans. *Novel Microb. Prod. Med. Agric. 1,* 161-169.

Bushley, K. E., Raja, R., Jaiswal, P., Cumbie, J. S., Nonogaki, M., Boyd, A. E., Owensby, C. A., Knaus, B. J., Elser, J., Miller, D., Di, Y., McPhail, K. L., Spatafora, J. W. (2013). The genome of Tolypocladium inflatum: evolution, organization, and expression of the cyclosporin biosynthetic gene cluster. *PLoS Genet.* **9**(6), e1003496.

Cao, L., Huang, J., Li, J. (2007). Fermentation conditions of Sinopodophyllum hexandrum endophytic fungus on production of podophyllotoxin. *Shipin Yu Fajiao Gongye* **33**(9), 28-32.

Chao, Y., Chen, J. S., Hunt, V. M.,Kuron, G. W., Karkas, J. D., Liou, R., Alberts, A. W. (1991). Lowering of plasma cholesterol levels in animals by lovastatin and simvastatin. *Eur. J. Clin. Pharmacol.* **40**(1), S11-14.

Cichewicz, R. H. (2010). Epigenome manipulation as a pathway to new natural product scaffolds and their congeners. *Nat. Prod. Rep.* **27**(1), 11-22.

Clayden, J., Moran, W. J., Edwards, P.J., LaPlante, Steven R. (2009). The Challenge of Atropisomerism in Drug Discovery. *Angew. Chem. Int. Ed.* 48, 6398 – 6401.

Coelmont, L., Kaptein, S., Paeshuyse, J., Vliegen, I., Dumont, J. M., Vuagniaux, G., Neyts, J. (2009). Debio 025, a cyclophilin binding molecule, is highly efficient in clearing hepatitis C virus (HCV) replicon-containing cells when used alone or in combination with specifically targeted antiviral therapy for HCV (STAT-C) inhibitors. *Antimicrob. Agents Chemother.* **53**(3), 967-976.

Cragg, G. M. and Newman, D. J. (2005). Biodiversity: a continuing source of novel drug leads. *Pure Appl. Chem.* **77**(1), 7-24.

Cragg, G. M. and Newman, D. J. (2013). Natural products: A continuing source of novel drug leads. *Biochim. Biophys. Acta* **1830**(6), 3670-3695.

Cueto, M., Jensen, P. R., Kauffman, C., Fenical, W., Lobkovsky, E., Clardy, J. (2001). Pestalone, a New Antibiotic Produced by a Marine Fungus in Response to Bacterial Challenge. *J. Nat. Prod.* **64**(11), 1444-1446.

Danishefsky, S. J., Masters, J. J., Young, W. B., Link, J. T., Snyder, L. B., Magee, T. V., Jung, D. K., Isaacs, R. C. A., Bornmann, W. G., Alaimo, C. A., Coburn, C. A., Di Grandi, M.J. (1996). Total Synthesis of Baccatin III and Taxol. *J. Am. Chem.Soc.* **118**(12), 2843-2859.

Debbab, A., Aly, A. H., Edrada-Ebel, R. A., Wray, V., Pretsch, A., Pescitelli, G., Kurtan, T., Proksch, P.(2012). New anthracene derivatives: structure elucidation and antimicrobial activity. *Eur. J. Org. Chem.* 1351−1359.

Debono, M., Turner, W.W., LaGrandeur, L., Burkhardt, F.J., Nissen, J.S., Nichols, K. K., Rodriguez, M.J., Zweifel, M. J.(1995). Semisynthetic Chemical Modification of the

References

Antifungal Lipopeptide Echinocandin B (ECB): Structure-Activity Studies of the Lipophilic and Geometric Parameters of Polyarylated Acyl Analogs of ECB. *J. Med. Chem* **38**(17), 3271-3281.

Dias, D.A., Urban, S. & Roessner, U. (2012). A Historical Overview of Natural Products in Drug Discovery. *Metabolites* **2**(2),303-336.

Elander, R. P. (2003). Industrial production of β-lactam antibiotics. *Appl. Microbiol. Biotechnol.* **61**(5-6), 385-392.

Elbarbry, F. A. and A. S. Shoker (2007). Therapeutic drug measurement of mycophenolic acid derivatives in transplant patients. *Clin. Biochem.* **40**(11), 752-764.

Endo, A., Kuroda, M., Tsujita, Y. (1976). ML-236A, ML-236B, and ML-236C, new inhibitors of cholesterogenesis produced by Penicillium citrinum. *J. Antibiot.* **29**(12), 1346-1348.

Eyberger, A.L., Dondapati, R., Porter, J. R.(2006). Endophyte Fungal Isolates from Podophyllum peltatum Produce Podophyllotoxin. *J. Nat. Prod.* **69**(8), 1121-1124.

Fabricant, D. S. and Farnsworth, N. R. (2001). The value of plants used in traditional medicine for drug discovery. *Environ. Health Perspect. Suppl.* **109**(1), 69-75.

Faeth, S. H. and William F. F. (2002). Fungal endophytes: common host plant symbionts but uncommon mutualists. *Integr. Comp. Biol.* **42**(2), 360-368.

Fleming, A. (1929). The antibacterial action of cultures of a Penicillium, with special reference to their use in the isolation of *B. influenzae*. *Br. J. Exp. Pathol.* **10**, 226-236.

Fleming, A. (1980). On the Antibacterial Action of Cultures of a Penicillium, with Special Reference to Their Use in the Isolation of B. influenzae. *Rev. Infect. Dis.* **2**(1), 129-139.

Gill, M., Gimenez, A., Jhingran, A. G., Smrdel, A. F. (1989). Dihydroanthracenones from *Dermocybe splendida* and related fungi. *Phytochemistry* **28**(10), 2647-2650.

Gould, I. M., David, M. Z., Esposito, S., Garau, J., Mazzei, T., Peters, G. (2012). New insights into meticillin-resistant *Staphylococcus aureus* (MRSA) pathogenesis, treatment and resistance. *Int. J. Antimicrob. Agents* **39**(2), 96-104.

Grove, J. F.,MacMillan, J., Mulholland, T. P. C., Rogers, M. A. T. (1952). Griseofulvin. IV. Structure. *J. Chem. Soc.* 3977-3987.

Guerram, M.,Jiang, Z-Z., Zhang, L-Y. (2012). Podophyllotoxin, a medicinal agent of plant origin: past, present and future. *Zhongguo Tianran Yaowu* **10**(3), 161-169.

Guo, B., Li, H., Zhang, L. (1998). Isolation of a fungus producing vinblastine. *Yunnan Daxue Xuebao, Ziran Kexueban* **20**(3), 214-215.

Hartley, S. E. and Gange, A. C. (2009). Impacts of plant symbiotic fungi on insect herbivores: mutualism in a multitrophic context. *Annu. Rev. Entomol.* **54**, 323-342.

Henrikson, J. C., Hoover, A. R., Joyner, P. M., Cichewicz, R. H. (2009). A chemical epigenetics approach for engineering the in situ biosynthesis of a cryptic natural product from *Aspergillus niger*. *Org. Biomol. Chem.* **7**(3), 435-438.

References

Ho, Y-S.,Duh, J-S.,Jeng, J-H., Wang, Y-J., Liang, Y-C., Lin, C-H.,Tseng, C-J.,Yu, C.-F.,Chen, R-J., Lin, J-K. (2001). Griseofulvin potentiates antitumorigenesis effects of nocodazole through induction of apoptosis and G2/M cell cycle arrest in human colorectal cancer cells. *Int. J. Cancer* **91**(3), 393-401.

Huber, F. M. and Gottlieb, D. (1968). The mechanism of action of griseofulvin. *Can. J. Microbiol.* **14**(2), 111-118.

Isaka, M., Jaturapat, A., Rukseree, K., Danwisetkanjana, K.,Tanticharoen, M., Thebtaranonth, Y. (2001). Phomoxanthones A and B, Novel Xanthone Dimers from the Endophytic Fungus Phomopsis Species. *J. Nat. Prod.* **64**(8), 1015-1018

Johnson, I. S., Armstrong, J. G., Gorman, M., Burnett, J. P. (1963). The Vinca Alkaloids: A new Class of Oncolytic Agents. *Cancer Res.* **23**, 1390-1427.

Kale, S. P., Milde, L., Trapp, M. K., Frisvad, J. C., Keller, N. P., Bok, J. W. (2008). Requirement of LaeA for secondary metabolism and sclerotial production in *Aspergillus flavus*. *Fungal Genet. Biol.* **45**(10), 1422-1429.

Kennedy, J. and Turner, G. (1996). δ-(L-α-Aminoadipyl)-L-cysteinyl-D-valine synthetase is a rate limiting enzyme for penicillin production in *Aspergillus nidulans*. *Mol. Gen. Genet.* **253**(1-2), 189-197.

Klayman, D. L. (1985). Qinghaosu (artemisinin): an antimalarial drug from China. *Science* **228**(4703), 1049-1055.

König, C. C., Scherlach, K., Schroeckh, V., Horn, F., Nietzsche, S.,Brakhage, A. A., Hertweck, C. (2013). Bacterium Induces Cryptic Meroterpenoid Pathway in the Pathogenic Fungus *Aspergillus fumigatus*. *ChemBioChem* **14**(8), 938-942.

Kosalková, K., García-Estrada, C., Ullán, R. V., Godio, R. P., Feltrer, R., Teijeira, F., Mauriz, E., Martín, J. F. (2009). The global regulator LaeA controls penicillin biosynthesis, pigmentation and sporulation, but not roquefortine C synthesis in *Penicillium chrysogenum*. *Biochimie* **91**(2), 214-225.

Kour, A., Shawl, A. S., Rehman, S., Sultan, P., Qazi, P. H., Suden, P., Khajuria, R. K., Verma, V. (2008). Isolation and identification of an endophytic strain of Fusarium oxysporum producing podophyllotoxin from *Juniperus recurva*. *World J. Microbiol. Biotechnol.* **24**(7), 1115-1121.

Kumaran, R. S., Muthumary, J., Kim, E-K., Hur, B-K. (2009). Production of taxol from Phyllosticta dioscoreae, a leaf spot fungus isolated from *Hibiscus rosa-sinensis*. *Biotechnol. Bioprocess Eng.* **14**(1), 76-83.

Kumaran, R. S., Muthumary, J., Hur, B. K. (2008). Taxol from *Phyllosticta citricarpa*, a leaf spot fungus of the angiosperm *Citrus medica*. *J. Biosci. Bioeng.* **106**(1), 103-106.

Kurobane, I., Vining, L.C., McInnes, A.G. (1979). Biosynthetic relationships among the secalonic acids. Isolation of emodin, endocrocin and secalonic acids from *Pyrenochaeta terrestris* and *Aspergillus aculeatus*. 32(12), 1256-1266.

Kusari, S., Lamshoeft, M., Spiteller, M. (2009). Aspergillus fumigatus Fresenius, an endophytic fungus from *Juniperus communis L. Horstmann* as a novel source of the anticancer pro-drug deoxypodophyllotoxin. *J. Appl. Microbiol.* **107**(3), 1019-1030.

References

Kusari, S., Zühlke, S., Spiteller, M. (2009). An Endophytic Fungus from *Camptotheca acuminata* That Produces Camptothecin and Analogues. *J. Nat. Prod.* **72**(1), 2-7.

Lanternier, F., Boutboul, D., Menotti, J., Chandesris, M. O., Sarfati, C., Mamzer-Bruneel, M. F., Calmus, Y., Mechai, F., Viard, J. P., Lecuit, M., Bougnoux, M. E., Lortholary, O. (2009). Microsporidiosis in solid organ transplant recipients: two Enterocytozoon bieneusi cases and review. Centre d'Infectiologie Necker-Pasteur, Service des Maladies Infectieuses et Tropicales, Hopital Necker-Enfants malades, Universite Paris Descartes, Paris, France: 83-88.

Lewington, S., Whitlock, G., Clarke, R., Sherliker, P., Emberson, J., Halsey, J., Qizilbash, N., Peto, R., Collins, R. (2007). Blood cholesterol and vascular mortality by age, sex, and blood pressure: a meta-analysis of individual data from 61 prospective studies with 55,000 vascular deaths. *Lancet.* **370**(9602), 1829-1839.

Liu, C. L. and Jiao, B. H. (2006). LC determination of Podophyllum lignans and flavonoids in *Podophyllum emodi Wall. var. chinesis Sprague*. *Chromatographia* **64**(9-10), 603-607.

Liu, K., Ding, X., Deng, B., Chen, W. (2009). Isolation and characterization of endophytic taxol-producing fungi from *Taxus chinensis*. *J. Ind. Microbiol. Biotechnol.* **36**(9), 1171-1177.

Liu, Y-Q., Yang, L., Tian, X. (2007). Podophyllotoxin: current perspectives. *Curr. Bioact. Compd.* **3**(1), 37-66.

Maiya, S., Grundmann, A., Li, S-M., Turner, G. (2006). The Fumitremorgin Gene Cluster of *Aspergillus fumigatus*: Identification of a Gene Encoding Brevianamide F Synthetase. *ChemBioChem* **7**(7), 1062-1069.

McCowen, M. C., Callender, M.E., Lawlis, J.F. (1951). Fumagillin (H-3), a New Antibiotic with Amebicidal Properties. *Science* **113**(2930), 202-203.

Min, C. and Wang, X. (2009). Isolation and identification of the 10-hydroxycamptothecin-producing endophytic fungi from Camptotheca acuminata Decne. *Xibei Zhiwu Xuebao* **29**(3), 614-617.

Mishra, B. B. and Tiwari, V. K. (2011). Natural products: An evolving role in future drug discovery. *Eur. J. Med. Chem.* **46**(10), 4769-4807.

Mita, T., Tanabe, K., Kita, K. (2009). Spread and evolution of Plasmodium falciparum drug resistance. *Int. J. Parasitol.* **58**(3), 201-209.

Nadeem, M., Ram, M., Alam, P., Ahmad, M. M., Mohammad, A., Al-Qurainy, F., Khan, S., Abdin, M. Z. (2012). Fusarium solani, P1, a new endophytic podophyllotoxin-producing fungus from roots of *Podophyllum hexandrum*. *Afr. J. Microbiol. Res.* **6**(10), 2493-2499.

Newman, D. J. and Cragg, G. M. (2012). Natural Products As Sources of New Drugs over the 30 Years from 1981 to 2010. *J. Nat. Prod.* **75**(3), 311-335.

Newton, G. G. and Abraham, E. P. (1955). Cephalosporin C, a new antibiotic containing sulphur and D-alpha-aminoadipic acid. *Nature* **175**(4456), 548.

Nicolaou, K. C., Simmons, N. L., Chen, J. S., Haste, N. M., Nizet, V. (1994). Total synthesis of taxol." *Nature* **367**(6464), 630-634.

References

Nützmann, H.-W., Reyes-Dominguez, Y., Scherlach, K., Schroeckh, V., Horn, F., Gacek, A.,Schumann, J., Hertweck, C., Strauss, J., Brakhage, A. A. (2011). ,Bacteria-induced natural product formation in the fungus Aspergillus nidulans requires Saga/Ada-mediated histone acetylation. *Proc. Natl. Acad. Sci. USA.* **108**(34), 14282-14287.

Oh, D.-C., Jensen, P. R., Kauffman, C. A., Fenical, W. (2005). ,Libertellenones A–D: Induction of cytotoxic diterpenoid biosynthesis by marine microbial competition., *Bioorg. Med. Chem.* **13**(17), 5267-5273.

Oh, D.-C., Kauffman, C. A., Jensen, P. R., Fenical, W. (2007). "Induced Production of Emericellamides A and B from the Marine-Derived Fungus *Emericella* sp. in Competing Co-culture." *J. Nat. Prod.* **70**(4), 515-520.

Panda, D., Rathinasamy, K., Santra, M. K., Wilson, L. (2005). Kinetic suppression of microtubule dynamic instability by griseofulvin: Implications for its possible use in the treatment of cancer. *Proc. Natl. Acad. Sci. USA.* **102**(28), 9878-9883.

Puri, S. C., Nazir, A., Chawla, R., Arora, R., Riyaz-ul-Hasan, S., Amna, T., Ahmed, B., Verma, V., Singh, S., Sagar, R., Sharma, A., Kumar, R.,Sharma, R. K., Qazi, G. N. (2006). The endophytic fungus *Trametes hirsuta* as a novel alternative source of podophyllotoxin and related aryl tetralin lignans. *J. Biotechnol.* **122**(4), 494-510.

Puri, S. C., Verma, V., Amna, T., Qazi, G. N., Spiteller, M. (2005). An Endophytic Fungus from Nothapodytes foetida that Produces Camptothecin. *J. Nat. Prod.* **68**(12), 1717-1719.

Ramesha, B. T., Amna, T., Ravikanth, G., Gunaga, R. P., Vasudeva, R., Ganeshaiah, K. N., Uma S. R., Khajuria, R. K., Puri, S. C., Qazi, G. N. (2008). Prospecting for Camptothecines from *Nothapodytes nimmoniana* in the Western Ghats, South India: identification of high-yielding sources of camptothecin and new families of camptothecines. *J. Chromatogr. Sci.* **46**(4), 362-368.

Rehman, S., Shawl, A. S., Verma, V., Kour, A., Athar, M., Andrabi, R., Sultan, P., Qazi, G. N. (2008). An endophytic Neurospora sp. from *Nothapodytes foetida* producing camptothecin. *Prikl. Biokhim. Mikrobiol.* **44**(2), 225-231.

Rodriguez, R. and Redman, R. (2008). More than 400 million years of evolution and some plants still can't make it on their own: plant stress tolerance via fungal symbiosis. *J. Exp. Bot.* **59**(5), 1109-1114.

Rönsberg, D., Debbab, A., Mándi, A., Vasylyeva, Vera., Böhler, P., Stork, B., Engelke, L., Hamacher, A., Sawadogo, R., Diederich, M., Wray, Victor., Lin, W., Kassack, M. U., Janiak, C., Scheu, S., Wesselborg, S., Kurtán, T., Aly, A. H., Proksch, P. (2013). Pro-Apoptotic and Immunostimulatory Tetrahydroxanthone Dimers from the Endophytic Fungus *Phomopsis longicolla*. *J. Org. Chem.*

Saikkonen, K., Faeth, S. H., Helander, M., Sullivan, T. J. (1998). Fungal Endophytes: A Continuum of Interactions with Host Plants. *Annu. Rev. Ecol. Evol. Syst.* **29**, 319-343.

Scherlach, K. and Hertweck, C. (2009). Triggering cryptic natural product biosynthesis in microorganisms. *Org. Biomol. Chem.* **7**(9), 1753-1760.

References

Scherlach, K., Nuetzmann, H-W., Schroeckh, V., Dahse, H-M., Brakhage, A. A., Hertweck, C. (2011). Cytotoxic Pheofungins from an Engineered Fungus Impaired in Posttranslational Protein Modification. *Angew. Chem., Int. Ed.* **50**(42), 9843-9847.

Schiff, P. B. and Horwitz, S. B. (1980). Taxol stabilizes microtubules in mouse fibroblast cells. *Proc. Natl. Acad. Sci. USA.* **77**(3), 1561-1565.

Schroeckh, V., Scherlach, K, Nutzmann, H.-W., Shelest, E., Schmidt-Heck, W., Schuemann, J., Martin, K., Hertweck, C., Brakhage, A. A. (2009). Intimate bacterial–fungal interaction triggers biosynthesis of archetypal polyketides in *Aspergillus nidulans*. *Proc. Natl. Acad. Sci. USA.* **106**(34), 14558-14563.

Shwab, E. K., Bok, J.W., Tribus, M., Galehr, J., Graessle, S., Keller, N.P. (2007). Histone deacetylase activity regulates chemical diversity in *Aspergillus*. *Eukaryot. Cell* **6**, 1656–1664.

Shweta, S., Zuehlke, S., Ramesha, B. T., Priti, V., Mohana, Kumar P., Ravikanth, G., Spiteller, M., Vasudeva, R., Uma, S. R. Endophytic fungal strains of *Fusarium solani*, from *Apodytes dimidiata E. Mey. ex Arn (Icacinaceae)* produce camptothecin, 10-hydroxycamptothecin and 9-methoxycamptothecin. *Phytochemistry* **71**(1), 117-122.

Soliman, S. S. M. and Raizada, M. N. (2013). Interactions between co-habitating fungi elicit synthesis of Taxol from an endophytic fungus in host Taxus plants. *Front. Fungi Their Interact.* **4**(Jan.), 3.

Srivastava, V., Negi, A. S., Kumar, J. K., Gupta, M. M., Khanuja, S. P. (2005). Plant-based anticancer molecules: A chemical and biological profile of some important leads. *Bioorg. Med. Chem.* **13**(21), 5892-5908.

Stefani, S., Chung, D. R., Lindsay, J. A., Friedrich, A. W., Kearns, A. M., Westh, H., MacKenzie, F.M. (2012). Meticillin-resistant Staphylococcus aureus (MRSA): global epidemiology and harmonisation of typing methods. *Int. J. Antimicrob. Agents* **39**(4), 273-282.

Strobel, G. and Daisy, B. (2003). Bioprospecting for microbial endophytes and their natural products. *Microbiol. Mol. Biol. Rev.* **67**(4), 491-502.

Strobel, G.,Daisy, B., Castillo, U., Harper, J. (2004). Natural Products from Endophytic Microorganisms. *J. Nat. Prod.* **67**(2), 257-268.

Strobel, G. A., Hess, W. M., Ford, E., Sidhu, R. S., Yang, X. (1996). Taxol from fungal endophytes and the issue of biodiversity. *J. Ind. Microbiol.* **17**(5-6), 417-423.

Sun, X., Guo, L.-D., Hyde, K.. (2011). Community composition of endophytic fungi in *Acer truncatum* and their role in decomposition. *Fungal Divers.* **47**(1), 85-95.

Survase, S. A., Kagliwal, L. D., Annapure, U. S., Singhal, R. S. (2011). Cyclosporin A — A review on fermentative production, downstream processing and pharmacological applications. *Biotech. Adv.* **29**(4), 418-435.

Szewczyk, E., Chiang, Y.-M., Oakley, C. E., Davidson, A. D., Wang, C. C., Oakley, B.R. (2008). Identification and characterization of the asperthecin gene cluster of Aspergillus nidulans. *Appl. Environ. Microbiol.* **74**(24), 7607-7612.

References

Vennerstrom, J. L., Arbe-Barnes, S., Brun, R., Charman, S.A., Chiu, F. C. K., Chollet, J., Dong, Y., Dorn, A.,Hunziker, D., Matile, H., McIntosh, K., Padmanilayam, M., Santo Tomas, J., Scheurer, C., Scorneaux, B., Tang, Y., Urwyler, H., Wittlin, S., Charman, W. N. (2004). Identification of an antimalarial synthetic trioxolane drug development candidate. *Nature* **430**(7002), 900-904.

Wall, M.E., Wani, M. C., Cook, C. E., Palmer, Keith H., McPhail, A. T., Sim, G. A. (1966). Plant Antitumor Agents. I. The Isolation and Structure of Camptothecin, a Novel Alkaloidal Leukemia and Tumor Inhibitor from *Camptotheca acuminata*. *J. Am. Chem. Soc.* **88**(16), 3888-3890.

Wani, M. C., Taylor, H. L., Wall, M. E., Coggon, P., McPhail, A. T. (1971). Plant antitumor agents. VI. Isolation and structure of taxol, a novel antileukemic and antitumor agent from *Taxus brevifolia*. *J. Am. Chem. Soc.* **93**(9), 2325-2327.

Wigley, D. B. (1995). Structure and mechanism of DNA topoisomerases. *Annu. Rev. Biophys. Biomol. Struct.* **24**, 185-208.

Williams, R. B., Henrikson, J. C., Hoover, A. R., Lee, A. E., Cichewicz, R. H. (2008). Epigenetic remodeling of the fungal secondary metabolome. *Org. Biomol. Chem.* **6**(11), 1895-1897.

Wink, M. (2008). Plant secondary metabolism: diversity, function and its evolution. *Nat. Prod. Commun.* **3**(8), 1205-1216.

Wongsrichanalai, C., Pickard, A. L., Wernsdorfer, W. H., Meshnick, S.R. (2002). Epidemiology of drug-resistant malaria. *Lancet Infect. Dis.* **2**(4), 209-218.

Yang, X., Zhang, L., Guo, B., Guo, S. (2004). Preliminary study of a vincristine-producing endophytic fungus isolated from leaves of *Catharanthus roseus*. *Zhongcaoyao* **35**(1), 79-81.

Zhang, L., Guo, B., Li, H., Zeng, S., Shao, H., Gu, S., Wei, R. (2000). Isolation of endophytic fungus of *Catharanthus roseus* and its fermentation to produce products of therapeutic. *Zhongcaoyao* **31**(11), 805-807.

Zhang, P., Zhou, P. P., Yu, L. J. (2009). An endophytic taxol-producing fungus from *Taxus media*, *Cladosporium cladosporioides* MD2. *Curr. Microbiol.* **59**(3), 227-232.

Zhao, J., Shan, T., Mou, Y., Zhou, L. (2011). Plant-derived bioactive compounds produced by endophytic fungi. *Mini Rev. Med. Chem.* **11**(2), 159-168.

Zuck, K. M., Shipley, S., Newman, D. J. (2011). Induced Production of N-Formyl Alkaloids from *Aspergillus fumigatus* by Co-culture with *Streptomyces peucetius*. *J. Nat. Prod.* **74**(7), 1653-1657.

List of Abbreviations

$[\alpha]^D$	specific rotation at the sodium D-line
br	broad signal
CD	Circular Dichroism
CH_2Cl_2	dichloromethane
$CDCl_3$	deuterated chloroform
$CHCl_3$	chloroform
CI	chemical ionization
cm	centimeter
COSY	correlation spectroscopy
d	doublet
DCM	dichloromethane
dd	doublet of doublet
DEPT	distortionless enhancement by polarization transfer
DMSO	dimethyl sulfoxide
DNA	deoxyribonucleic acid
ED	effective dose
EI	electron impact ionization
ESI	electrospray ionization
et al.	et altera (and others)
EtOAc	ethyl acetate
eV	electronvolt
FAB	fast atom bombardment
g	gram
HMBC	heteronuclear multiple bond connectivity
HMQC	heteronuclear multiple quantum coherence
H_2O	water
HPLC	high performance liquid chromatography
H_3PO_4	phosphoric acid
hr	hour
HR-MS	high resolution mass spectrometry
Hz	Herz
IZ	inhibition zone
L	liter
LC	liquid chromatography
LC/MS	liquid chromatography-mass spectrometery
m	multiplet
M	molar
MeOD	deuterated methanol
MeOH	methanol
mg	milligram
MHz	mega Herz
min	minute

Abbreviations

mL	milliliter
mm	millimeter
MS	mass spectrometry
MTT	microculture tetrazolium assay
m/z	mass per charge
µg	microgram
µL	microliter
µM	micromol
NaCl	sodium chloride
ng	nanogram
NMR	nuclear magnetic resonance
NOE	nuclear Overhauser effect
NOESY	nuclear Overhauser and exchange spectroscopy
PCR	polymerase chain reaction
ppm	parts per million
q	quartet
ROESY	rotating frame overhauser enhancement spectroscopy
RP 18	reversed phase C 18
s	singlet
sp.	species (singular)
t	triplet
TFA	trifluoroacetic acid
THF	tetrahydrofuran
TLC	thin layer chromatography
UV	ultra-violet
VLC	vacuum liquid chromatography

Research Contribution

Publications

Antonius R.B. Ola, Dhana Thomy, Daowan Lai, Heike Brötz-Oesterhelt, Peter Proksch. Inducing Secondary Metabolite Production of the Endophytic Fungus *Fusarium tricinctum* through Co-culture with *Bacillus subtilis*. *J. Nat. Prod.*, 2013, 76, 2094-2099

Antonius R.B. Ola, Amal H. Aly, Ilka Zerfass, Alexandra Hamacher, Attila Mandi, Matthias Kassack, Heike Brötz-Oesterhelt, Tibor Kurtan, Peter Proksch, Abdessamad Debbab. Absolute Configuration and Antibiotic Activity of Neosartorin from the Endophytic Fungus *Aspergillus fumigatiaffinis*. Tetrahedron Letter 2013 (**In Press**, available on line 26 december 2013).

Antonius R.B. Ola, Heike Brötz-Oesterhelt, Abdessamad Debbab, Peter Proksch and Amal H. Aly. Dihydroanthracenone Metabolites from the Endophytic Fungus *Diaporthe melonis* Isolated from *Anonna squamosa*. Tetrahedron Letters (2013 submitted).

Declaration of Academic Honesty/Erklärung

Hiermit erkläre ich ehrenwörtlich, dass ich die vorliegende Dissertation mit dem Titel „Sekundärstoffe aus endophytischen Pilzen - Ansätze zur Aktivierung stiller Biosynthesewege, Strukturaufklärung und Bioaktivität" selbst angefertigt habe. Außer den angegebenen Quellen und Hilfsmitteln wurden keine weiteren verwendet. Diese Dissertation wurde weder in gleicher noch in abgewandelter Form in einem anderen Prüfungsverfahren vorgelegt. Weiterhin erkläre ich, dass ich früher weder akademische Grade erworben habe, noch dies versucht habe.

Düsseldorf, den 10.01.2014

Antonius R B Ola

Acknowledgment

It is a pleasure to find the chance and the moment to thank and to acknowledge all the people that directly or indirectly supported and involved during this doctoral study until the completion of this doctoral thesis.

At the first place, I would like to express my special thanks and gratitude to Prof. Dr. rer. nat. Peter Proksch for giving me the opportunity to undertake my doctoral study at the institute, as well as for all his scientific valuable suggestions, his fruitful discussions, his unforgettable support and for the excellent working conditions and facilities at the Institut für Pharmazeutische Biologie und Biotechnologie, Heinrich-Heine-Universität, Düsseldorf. The hard sculpture will create a beautiful art and I would like to acknowledge from the deepest for shaping all the scientific aspect and everything during my study.

I am deeply indebted to Dr. Amal Hassan for all her scientific and constructive advises, NMR courses, sharing her expertise in NMR data interpretation and preparations of the manuscripts for the publications.

I would like to express my special appreciation for Dr. Daowan Lai for sharing the scientific knowledge and the genius NMR interpretation for structure elucidations and a warm friendship during the study.

I would also like to thank to Prof. Dr. H. Brotz-Oesterhelt for her kind and warm support in the microbiology part: coculture research projects and antibiotic assays, for all the advice and help especially for the coculture project and the publications.

My deep thanks and gratitude to Prof. Dr. rer. nat. Matthias Kassack, Institut für Pharmazeutische und Medizinische Chemie, Heinrich-Heine University, Düsseldorf, for his recommendation, conducting cytotoxicity assays and for estimating my PhD study as second referee.

I am deeply grateful to Dr. Abdessamad Debbab for all the sharing of techniques and knowledge, the valuable help and scientific advices, NMR courses, together with sharing his expertise in NMR data interpretations.

I would like to thank Prof. Dr. Claus Passreiter for his support, help and safety matters during the study.

Acknowledgment

I would like also to express my deep thanks to Mrs. Waltraud Schlag, Mrs. Simone Miljanovic, Mrs. Katja Friedrich and Mrs. Claudia Eckelskemper together Mr. Dieter Jansen and Frau Eva Müller for their hospitality, supports and helps during the study. Thank you so much.

This study will not finish and complete without the help, support, sharing of knowledge, smile and laughing from all friends, family and everybody whom I met or maybe not. At this special moment, I would like to express my special thanks:

- Dr. rer. nat. Robert Bara for being a big Brother since the start of my study. We have walked together in the bitterness and joys throughout the years in Düsseldorf. The time we spent together was so wonderful. It is a pity that we live so far away again in Indonesia. Thanks for being my second wife in the Lab. Every chemicals and glass are always ready to be used because of ROBY. Once more thank you very much dear Robot Bara, I will remember you always!
- Mr. Andreas Marmann and Ms. Lena Hammerschmidt for the early work in fungi group. Thank for showing me how to fight against all the fungi.
- Thanks to the old friends that have been living Dr. David Rönsberg, Dr. Bartoz Lipovics, Dr. Yaming Zhou, Dr. Weam Ibrahim, Dr. Sherif El Sayed, Mr. Mustafa El Amrani, kang Asep Supriadin, Dr. Jiang Ping Wang and Nadine.
- Dr. Chong Dat Pharm, Ingo Kolb and Mr. Hendrik Niemann for the time being together, PingPong and for your nice friendship.
- Dr. Ilka Zerfass, Heike Goldbach-Gecke and Ms. Dhana Tommy for helping us throughout the coculture project and antibiotic assays. Thank you so much.
- Catalina Peres Hemphil, Imke Form and Mi-Young Chung for all your help and nice conversations.
- I have enjoyed the time here with all of you George Daletos, Sergi Akone, Amin Mokhlesi, Mousa Al Tarabee, Shuai Lew, Yaolong Lew, Jens and Steffi, Fatima Kabbaj and Artha Kuci.
- Tatyana Medvedeva and Nihal Aktas for being nice friends and warm for the small lab
- Special thanks also to the Chinese friends: Huiqin Chen, Yang Liu, Fang lü!
- Special thanks also to the visiting professors: Prof. Binggui Wang, Prof. Haofu Dai, Prof. Wenhan Lin and Prof. Veneta Ivanova for all scientific discussion and help.
- Terima kasih yang terdalam buat mba Rini dan mas Ruddy beserta kang Asep supriadin atas kebersamaannya, senang bisa berjumpa dan berkenalan dengan kalian.

- I would like also to express my special thanks for our Deutsche family member: Gitta and Ruddy Lammert. Terima kasih banyak!
- Special appreciation for Indonesian Catholic Community in Düsseldorf (KoKid) for every nice moment we have until the baptism for our Alexander Pedro.
- Special thanks for our fathers Polycarpus Ungarn and Cyril Binsasi for their spirit and wisdom words and preaches during our stay in Düsseldorf.

This work during the study was foremost devoted to my beloved ones Maria Renny Praptiwi, Maria Grace M. Aussiola and Alexander Pedro Ola.

i want morebooks!

Buy your books fast and straightforward online - at one of world's fastest growing online book stores! Environmentally sound due to Print-on-Demand technologies.

Buy your books online at
www.get-morebooks.com

Kaufen Sie Ihre Bücher schnell und unkompliziert online – auf einer der am schnellsten wachsenden Buchhandelsplattformen weltweit! Dank Print-On-Demand umwelt- und ressourcenschonend produziert.

Bücher schneller online kaufen
www.morebooks.de

VDM Verlagsservicegesellschaft mbH
Heinrich-Böcking-Str. 6-8
D - 66121 Saarbrücken

Telefon: +49 681 3720 174
Telefax: +49 681 3720 1749

info@vdm-vsg.de
www.vdm-vsg.de

Printed by Books on Demand GmbH, Norderstedt / Germany